THE STRATEGIC BALANCE
IN THE MEDITERRANEAN

THE STRATEGIC BALANCE
IN THE MEDITERRANEAN

Jesse W. Lewis, Jr.
With a foreword by Elmo R. Zumwalt, Jr.

American Enterprise Institute for Public Policy Research
Washington, D. C.

Jesse W. Lewis, Jr., is political/military affairs officer at the
American Embassy in Saudi Arabia.

355.03301
L 58s
97056
may 1976

ISBN 0-8447-3197-8

Foreign Affairs Study 29, March 1976

Library of Congress Catalog Card No. 75-37446

Printed in the United States of America

To my mother,

to the memory of my father,

and

to all who have gone to sea in the Mediterranean

CONTENTS

LIST OF MAPS

FOREWORD

I commend highly to any student of political/military affairs or to any party interested in the strategic balance around the world this fine piece of research by Jesse W. Lewis, Jr. While neither I nor any other person would agree with everything written in any treatise regarding such a complicated subject as this, it is my view, as one who has also studied and written on the subject, that Mr. Lewis has made a major contribution with this scholarly work.

No one who reads this treatise can fail to be impressed by the dramatic growth both of Soviet naval capability and of the total power the Soviet Union can bring to bear in the Mediterranean, directly and through allies. No one who reads it can fail to be concerned by the deteriorating capabilities of the United States to influence events in the Mediterranean, partly as a result of the reduction in its own military capabilities and partly as a result of the weakening of alliance ties in that area.

It is an important challenge to all thoughtful students of this work to seek to go beyond it in the important task of guiding the destiny of the United States in this area in the years ahead.

ELMO R. ZUMWALT, JR.
Admiral, USN (Retired)

PREFACE

This is essentially a survey, a kind of annotated catalogue of the military presence in the Mediterranean Basin, combined with an outline of the active and dormant disputes that could alter the balance of power in the Mediterranean and cause the weapons listed here to be used.

The focus of this study is on the superpower presence—that of the United States and Soviet Union—though the armed strengths of the Mediterranean's fifteen continental and two island countries, and the British presence, which has made a significant contribution to Western security interests in the Mediterranean, are also considered. However, no meaningful examination of these subjects can exclude a discussion of the Mediterranean Sea itself and some of the basic elements of sea power in the twentieth century.

First a definition of terms: the continental countries that comprise the strictly geographical Mediterranean world are, starting at Gibraltar and moving clockwise, Spain, France, Italy, Yugoslavia, Albania, Greece, Turkey, Syria, Lebanon, Israel, Egypt, Libya, Tunisia, Algeria, and Morocco. The two island countries are Malta and Cyprus. There is also what might be called the geopolitical Mediterranean, which includes a number of Middle Eastern countries that do not have Mediterranean coastlines but are inextricably linked to the area by geography, oil, and politics; they are Jordan, the Sudan, the two Yemens, Iraq, Iran, Kuwait, Saudi Arabia, and the smaller Arab states of the Persian Gulf. Portugal is also part of this expanded Mediterranean world.

A word on sources. Most of the statistical data and other information on arms was taken from the *1974–75* and *1975–76 Military Balance* published by the International Institute for Strategic Studies and the authoritative *Jane's* volumes, *Fighting Ships, All the World's*

Aircraft, and *Weapon Systems.* I am greatly indebted to these publications and their editors, particularly Captain John E. Moore, RN (Ret.), of *Jane's Fighting Ships.*

This study could not have been undertaken or completed without the generous assistance of the American Enterprise Institute for Public Policy Research, which awarded me a research fellowship in 1972. I also want to thank the *Washington Post* for which I was working as Middle East correspondent when this study was conceived, and to express gratitude for a sabbatical leave of absence that allowed me to begin the work.

During the past six years, as a journalist or researcher I have visited every Mediterranean country except Albania. During my trips I interviewed several hundred people whose employment brings them into daily contact with the Mediterranean. They included government officials, naval officers, ship captains and shipping executives, historians, and journalists. All provided me with invaluable information and advice. They are too numerous to mention individually, and many who helped did so with the understanding that they would remain anonymous. They know who they are and they also know that they have my gratitude.

However, I want to individually thank John W. Warner, who was secretary of the navy when I started this study and who was instrumental in arranging for me to visit U.S. Navy units at sea and ashore in the Mediterranean. My sincere thanks go also to several former Sixth Fleet commanders: Admiral Isaac C. Kidd, Jr., Vice Admiral Daniel J. Murphy, and Vice Admiral Gerald E. Miller (Ret.), and to Vice Admiral Pierre N. Charbonnet, Jr., who was CTF-67 during much of my research. I also want to thank Norman Polmar, the editor of the United States section of *Jane's Fighting Ships,* for reading my manuscript and making constructive comments.

My colleagues at AEI also deserve special mention, particularly Robert J. Pranger and Dale R. Tahtinen, respectively the director and assistant director of foreign and defense policy studies; researchers John Lenczowski, Peter Chaffetz, Larry Kahn, and Craig N. Moore; and typists Iris McPherson, Lynn Balthaser, Madelyn Jones, and Karyn Pritchard.

I, of course, am responsible for all the conclusions and errors.

Washington, D. C.
June 1975

1
INTRODUCTION

The Mediterranean is the Helen among oceans. Like her it was desired by all who saw it . . . fought over not for ten, but for two thousand years, and captured by the boldest. Today before our eyes it is fought over anew.[1]

In many ways the Mediterranean is a barometer of the world's political climate. This is more true today than at any other time in modern history because developments in the seventeen Mediterranean countries and the state of relations among them often mirror the state of relations between the two superpowers, if only because the super-powers have an overwhelming strategic stake in the Mediterranean.

The energy crisis illustrates this point. The price and availability of petroleum is affected by the Arab-Israeli conflict whose major protagonists, Israel, Egypt, and Syria, are Mediterranean countries and whose major arms suppliers are the United States and the Soviet Union. The opening of the Suez Canal is another example of this equation in operation.

It is almost as if the United States and the Soviet Union had chosen the Mediterranean as an arena for contest: as a testing platform for their political ideologies, their economics, and their weapons. And now that the United States and the Soviet Union have made a cautious, but apparently determined effort to ease world tensions, the Mediterranean will inevitably serve as a laboratory for measuring détente.

However, the Mediterranean is far more than a litmus of the world's temper. It is the geographical, political, military, and economic—in a word, strategic—junction of three continents and even

---

[1] Emil Ludwig, *The Mediterranean: Saga of a Sea* (London: Hamish Hamilton, 1943).

1

more cultures—Southern Europe, the Balkan Peninsula, the Middle East, and North Africa. A staggering number of conflicting interests are packed around this small area. Most of these interests are vital, and many of the conflicts have hazardously low flash points. Taken together they form a complex constellation of challenges, opportunities, and risks, especially for the United States.

Yet, in the organization of American foreign policy, the Mediterranean, in my judgment, receives only piecemeal, ad hoc attention. There is no Mediterranean consciousness in the United States. Generally speaking, this region is considered as a whole only during crises, when it is often too late to prevent damage to American interests. As a political unit it may be too complex, too unwieldy to warrant, for instance, an assistant secretary of state for Mediterranean affairs. But it does deserve closer, more regular scrutiny.

What is at stake for the superpowers? The United States has a higher density of vital interests in the Mediterranean Basin than any other area of the world except the Americas. These interests range from protecting American-owned oil companies in the Middle East and North Africa and the contribution that they make to the U.S. balance of payments to keeping that oil steadily flowing to Western Europe and Japan, which could not survive without it; to maintaining the presence of the Sixth Fleet along NATO's southern flank as a credible military force and as a political-strategic counterweight to the growing influence of the Soviet Union in the eastern and southern Mediterranean; to preventing the Arab-Israeli dispute from boiling over into a wider conflict.

For the Soviet Union, the Mediterranean is the politically unreliable path to its back door, the Black Sea, and includes the major industrial and agricultural regions that the Black Sea borders. It is the "warm water" that has been the object of Russian foreign policy since the reign of Catherine the Great. The Soviet Union tried unsuccessfully to insert itself in the Mediterranean after World War II by encouraging a civil war in Greece and putting diplomatic pressure on Turkey. Then in 1955 the Soviet Union dramatically leaped over the "northern tier" by combination of carrot and cannon tactics—but this time offering the cannon as well. In 1955, Russia concluded an arms deal with Egypt that widened into an unprecedented political, military, and economic investment in the Mediterranean. It built the Aswan Dam in Egypt. It later sent its own pilots and missilemen to defend Egyptian airspace.

Though the Russian presence in Egypt suffered a setback in 1972 when Soviet advisors left, the Egyptian armed forces still rely on

2

Soviet arms and equipment, and Russian warships still use Egyptian port facilities. To compensate for the decline of its influence in Cairo, the Soviet Union made an even heavier investment in Syria, supplying it with weapons that have never before been deployed outside the Warsaw Pact area.

In the last ten years, Turkey has received substantial amounts of Soviet economic aid. Even pro-Western Iran receives both military and economic assistance from the Soviet Union. The armed forces of Lebanon, Libya, Algeria, and Morocco also have Soviet military hardware. The most powerful Russian naval force outside Soviet territorial waters is in the Mediterranean, the only sea where the Soviet Union keeps its navy constantly within striking distance of the American surface fleet and where it may possess some marginal ability against the Polaris/Poseidon submarine.

Twenty years ago, the United States could count on using air bases in North Africa during a crisis in the Mediterranean—not to speak of facilities in Europe. Today, this country has access to a single small naval station and communications facility in North Africa, in Morocco, and during the 1973 Arab-Israeli war American bases in Mediterranean Europe could not be used by American transports carrying emergency arms to Israel. Even Britain, the United States's closest ally, only reluctantly granted facilities at Akrotiri, one of its sovereign bases in Cyprus, to American U-2 aircraft to monitor the 1970 Suez Canal cease-fire agreement. This reversal is illustrated even more sharply by the fate of the American-built Wheelus Air Force Base outside Tripoli, Libya: Russian troops today train Libyan soldiers to use Soviet military equipment where, until 1970—but five years ago—American servicemen were supervising training with American equipment.

Today, the presence of American military forces and the facilities they use in the Mediterranean is under review—in Portugal, Greece, and Turkey. And restricted U.S. base rights are being renegotiated in Spain. This development is not only a reaction to a particular American policy, as in Cyprus, for example. Nor for that matter is the review the result of any success of Russian policy, though the Soviet Union has dramatically increased its naval presence in the Mediterranean. Rather, many nations are reevaluating their relationship with the United States in the wake of the Vietnam and Cambodia experiences, and at the same time the United States is reassessing its own place in the world. Nowhere is the outcome of this process more crucial to vital American interests than in the Mediterranean. The Mediterranean is a vast political echo chamber where developments in any one country—and many events in countries outside the region—are reverberated and intensified, often exploding with violence that in turn is felt in other parts of the globe. To say the same about other oceans and seas would be to miss the unique character of the Mediterranean—its compactness, its water, its widely different stages of economic and social development, its competing political systems, its instability and potential for violence, and its strategic importance to the Mediterranean countries themselves as well as to the United States and the Soviet Union.

Today, the Mediterranean—birthplace, cradle, and graveyard of civilizations—has the potential to engulf the world in a nuclear war that would mean the end of civilization as it is now known.

2
THE SEA AND THE IMPORTANCE OF NAVAL POWER

The sea is the primary platform of superpower presence in the Mediterranean. Let us begin, then, with a discussion of the sea itself and the importance of naval power.

Geography of the Mediterranean

Two of the most fundamental characteristics of the Mediterranean Sea are its size and its shape. Look at the sea on a world map. It is small, almost infinitesimal compared with other major bodies of water. Its area is 969,100 square miles, or, if you include the Black Sea, about 1,158,300 square miles.[1] It would take an American aircraft carrier making 25 knots (nautical miles per hour) three days and nine hours to cover the Mediterranean's length of 2,034 miles measured

---

[1] The question of whether the Black Sea is part of the Mediterranean is complex and involves more than a geographer's judgment. There is no general agreement on the answer. Some sources include the Black Sea as part of the Mediterranean as a matter of course, while others exclude it. The 1929 edition of *Encyclopedia Britannica*, for example, includes the Black Sea in a table showing the area of the Mediterranean. The 1955 edition left it up to the reader by saying: "Its total area, if one includes the Black Sea, approaches 1,158,300 sq. mi." However, the 1965 edition of *Britannica* put it this way: "Its area, including the Sea of Marmara but excluding the Black Sea, is close to 970,000 sq. mi." The basic geography of the Mediterranean did not change in the intervening years but the power and policy of the Soviet Union did. The central point is political, not geographic. If the Black Sea is part of the Mediterranean, then the Soviet Union, Bulgaria, and Rumania are Mediterranean in the political sense as well. The Montreux Convention that governs the use of the Turkish Straits connecting the Black Sea with the Mediterranean already gives these countries special rights of transit but denies unregulated access for warships. The broader aspects of this question will be discussed later in the study.

Table 1

COASTLINE MEASUREMENTS AND TERRITORIAL SEA LIMITS OF MEDITERRANEAN COUNTRIES

Country	Nautical Miles in Mediterranean [a]	Territorial Sea [b]
Albania	155	12
Algeria	596	12
Cyprus	290	12
Egypt	538	12
(Red Sea 769)		
France	266	12
Corsica 225		
(Atlantic and North Sea 882)		
Greece	1,645	6
Mainland and fringing islands 1,210		
Crete 340		
Rhodes 95		
Israel	120	6
(Gulf of Aqaba 4)		
Italy	2,451	6
Peninsular West Coast 690		
Peninsular East Coast 852		
Sicily 461 Sardinia 408		
Elba 40		
Lebanon	105	6
Libya	910	12
Malta, including Gozo	50	6
Monaco	3	3
Morocco	190	12
(Atlantic 705)		
Spain	658	6
Balearics 261		
(Atlantic 565) (Canary Islands 544)		
Syria	82	12
Tunisia	555	12
Turkey	974	6/12[c]
Sea of Marmara 239 (Black Sea 708)		
Yugoslavia	426	10

[a] Figures are taken from U.S. Department of State, Office of Strategic and Functional Research, Bureau of Intelligence and Research, "Sovereignty of the Sea," Geographic Bulletin No. 3, October 1969. (Distances given in the tabulation above represent the extent of each political area's coastline which "faces the sea" exclusive of detailed irregularities. Measurement was effected by swinging a divider over 10-mile intervals on maps at a scale of 1:1,000,000, the largest for which there is complete world coverage. Figures in parentheses refer to non-Mediterranean coastlines.)

[b] Department of State, The Geographer, Office of the Geographer, Bureau of Intelligence and Research, "Limits in the Sea No. 36, National Claims to Maritime Jurisdiction," revised 1 April 1974.

[c] The 12 miles applies in the Black Sea as a matter of reciprocity with other Black Sea powers. Ibid., p. 119.

from the Strait of Gibraltar to Beirut, Lebanon.[2] Because of the Mediterranean's irregular shape, its north-south measurement varies. A Soviet guided missile cruiser of the Kresta II class coming from the Black Sea has a trip of 814 miles from the point where it enters the Turkish Straits to Port Said, Egypt, the entrance to the Suez Canal. Such a ship cruising at 25 knots in open water would cover that distance in about 38 hours, allowing for slower speeds in the two congested channels of the Turkish Straits, the Bosporus and the Dardanelles.[3] A tanker carrying crude oil from Bejaia, a major oil terminal on Algeria's Mediterranean coast, to the refinery complex at Port-de-Bouc, near Marseille, France, can shuttle the 410 miles in just over 20 hours at a speed of 20 knots.

Compared with the Atlantic's area of 31,529,000 square miles, the Pacific's 63,985,000 square miles, or the total estimated 139,670,000 square miles of water that covers the globe, the Mediterranean is only a drop in the bucket. Some of the distances between ports in other oceans sharpen this contrast. From San Francisco to Yokohama, Japan, is 4,536 miles, and it would take a ship travelling at 25 knots seven days, 14 hours to make the journey. A freighter making the 11,977-mile run from Norfolk, Virginia, to Banda Mashur, an Iranian port on the Persian Gulf, averaging 20 knots would be at sea 18 days, 18 hours!

Another feature of the Mediterranean is that it is virtually enclosed and, though its coastline is irregular, the sea itself is rather neatly divided into two distinct basins. The only corridors to and from the larger oceans are the Strait of Gibraltar and the Suez Canal. (The Turkish Straits—the Dardanelles, the Sea of Marmara, and the Bosporus—connect the Mediterranean with a smaller, even more enclosed body of water, the Black Sea.) The Strait of Gibraltar, straddled by Spain and Morocco, is only 8 miles wide and is one of the most congested waterways in the world. When the Suez Canal was closed between 1967 and 1975 all traffic entering or leaving the

---

[2] A nautical mile is 6,076.115 feet, compared with 5,280 feet for a statute mile. A ship travelling at a speed of 10 knots, that is, 10 nautical miles an hour, would cover a distance of twenty nautical miles in two hours. Current, tide, wind, and sea conditions may alter this calculation. All distances are in nautical miles and were taken from Department of the Navy, Oceanographic Office, *Distances between Ports, 1965*, Pub. No. 151.

[3] In reality such a trip for a Soviet warship through the Turkish Straits would take much longer because the Montreux Convention requires Black Sea powers to give at least eight days notice to Turkey of intention to transit the straits. As a member of NATO, Turkey regularly passes on the notice to the alliance's military command which in turn passes it on to Allied navies in the Mediterranean.

Mediterranean passed through Gibraltar. About 200 miles inside this strait the sea balloons out to the north along the coasts of Spain, France, and Italy. To the south, the coasts of Morocco, Algeria, and Tunisia are relatively straight.

The western basin of the Mediterranean can be conveniently divided into three sub-basins or seas. The waters between the Balearic group of islands—Mallorca, Ibiza, and Menorca—and the coast of Spain are known as the Balearic Sea. The Ligurian Sea is between the Balearics and the coasts of France and Italy. The triangular area of water formed by the northern coast of Sicily, eastern Corsica and Sardinia, and western Italy is called the Tyrrhenian Sea.

In order to reach the eastern basin, ships must pass through the Strait of Sicily or the Strait of Messina. In the Strait of Sicily, the Mediterranean narrows to 83 miles and the African and European continental shelves extend far out into the sea to form underwater plateaus, with some areas as shallow as 36 feet. This reduces the navigable portion of the strait for large ships and submarines. Once on the other side, the sea bulges out again, this time to the south, forming a great bowl along eastern Tunisia, Libya, and Egypt. Curving back up and around Israel, Lebanon, and Syria, the Mediterranean envelops Cyprus and runs alongside Turkey until it reaches the island-dotted Aegean Sea, another subsidiary. After passing western Greece, the Mediterranean becomes an easily definable sea-within-a-sea, the slender Adriatic, which stretches past Albania, Yugoslavia's Dalmatian Coast, and back down the eastern frontier of Italy. The V-shaped body of water between western Greece and southern Italy is known to mariners as the Ionian Sea.

Choke Points. The Strait of Sicily, the Strait of Gilbraltar, and a few other channels in the Mediterranean are called "choke points" by naval strategists. These waterways are narrow and the concentration of ships in a relatively small patch of sea as they pass through makes them easy to monitor or attack. Choke points also can be mined or sown with underwater sensing devices linked to computers which can monitor the movements of submarines.

The most critical choke points for naval operations are the Strait of Gibraltar, the Strait of Sicily, and the Turkish Straits, which control access to and from the Black Sea. If Gibraltar, the Turkish Straits, and the Suez Canal were closed to shipping, any vessels inside the Mediterranean would be trapped and ships outside could not get in. If the Strait of Sicily were obstructed, ships caught in either basin would have to stay there. Some of the other choke points are the approaches to the Aegean Sea west and east of Crete; the Strait of Messina, a 2-mile-wide channel between Sicily and Italy; and the Strait of Otranto, the 41-mile-wide corridor to the Adriatic Sea between Albania and Italy.

The Strait of Sicily and the restricted waters of the central Mediterranean were the focus of much fighting during World War II. German and Italian convoys had to transit the area to replenish Rommel's troops in North Africa. The same area was part of Britain's lifeline to its forces in Malta and Egypt. This area between Sicily, Malta, and the Tunisian coast is still an important dividing line in the Mediterranean. The Sixth Fleet normally has two aircraft carriers in the sea, one west of this line, the other east. The ease of interdicting trans-Mediterranean shipping may be one reason why the Soviet Union has concentrated its forces in the eastern basin of the sea where its supply lines are shorter and more secure.

At several places in addition to the Strait of Sicily the sea floor rises steeply in international waters to make large patches shallow enough for ocean-going ships to anchor. The Soviet fleet uses these areas as "bases at sea." The ones it used most often are: Alboran Island, just east of the Strait of Gibraltar; Hurd Bank, south of Malta; the Gulf of Hammamet, east of Tunis; Kithira, south of Kithira Island and east of Crete; and Cape Andreas, off the eastern tip of Cyprus. Their location near the choke points enables Soviet ships anchored at these spots to monitor Western naval activities, to take on food and fuel, and to ease the rigors of underway watch schedules for Russian sailors. The anchorages also compensate for the absence of Russian naval bases in the northern and western Mediterranean.

Antisubmarine Warfare. The steep underwater gradients which rise to form the plateaus combine with "thermal layers" to make the waters of the Mediterranean among the most difficult in the world in which to locate submerged submarines. Sound waves—both the noise produced by submarines' propellors, engines, and other machinery, and signals generated by active sonar—are distorted as they bounce off the walls of underwater canyons or pass through water whose temperature varies with depth. The differences in temperature are most pronounced during summer months when the sea is covered with a sun-heated layer that can reach as deep as 180 feet. A submarine cruising at, say, 250 feet below the surface will escape detection by surface sonar. In August, for example, the surface temperature of 76 degrees farenheit extends down to 120 feet. A surface ship with passive sonar, that is, sonar that just listens for a submarine's noises, would not be able to get a clear signal from a nuclear-powered submarine in the lower layer where the temperature is 56 degrees. The commander of an American nuclear-powered hunter-killer sub told me: "You can put the whole fleet out there in the summertime and it wouldn't do any good unless you have VDS." [4]

The high salinity of the Mediterranean also affects the transmission of sound waves, all the more so now that the Aswan High Dam restricts the fresh water flow of the Nile River. The higher the salinity of water, the faster sound waves travel through it.

A word about how sonar works. As mentioned earlier, there are two basic types of sonar—passive and active. An active sonar system emits electronically produced noise, or "pings," which is what the noise sounds like if you are in a submarine that is the target of an active sonar search. These pings are so powerful that a person swimming below the surface can suffer a fatal concussion if he is in the path of an active sonar emission. Detection of submerged objects works on the principle of the speed of sound through water. Sonar operators can compensate for the variations caused by temperature (though not the pronounced differences between thermal layers) and salinity and can calculate how much time it takes for a sound to reach their sensors or to bounce back if they are using active sonar. A trained sonar operator using a modern system can tell the speed of a vessel, its range, depth, course, number of propellors, and number of blades on the propellors—from which information he can identify what kind of ship he is tuned in on.

On any given day, there are more than 2,500 merchant ships of over 1,000 tons underway in the Mediterranean. There are another 5,000 or so smaller fishing, coastal, and pleasure craft making their way through the blue water. (The cadence of their propellors churning the water is called "high ambient noise" by sonar operators. It is another sound they must cope with when trying to locate submarines.) These vessels travel between Marseille and Haifa, New York and Barcelona, Beirut and Piraeus, Galveston and Istanbul, Rotterdam and Tripoli. The reopened Suez Canal will bring even more ships into the Mediterranean.

---

[4] VDS—for variable depth sonar—is one method that can overcome the effect of thermal layers. Some ships of the French navy, which has the most highly developed VDS system, have large pods weighing 10,000 pounds mounted on their fantails that are lowered overboard with several hundred feet of cable. These pods, which resemble spherical fish with moveable fins to give them directional control, are trailed behind the antisubmarine warfare (ASW) ships and can emit and listen to underwater sounds below thermal layers.

The present high volume and complex pattern of sea traffic is another feature of the Mediterranean that helps to measure its importance both to the countries whose shores it washes and to other parts of the world. It does not take much imagination to grasp the impact of any sustained interruption of shipping in the Mediterranean, particularly of oil and other vital supplies. It is true that supertankers shuttle between the Persian Gulf and Western Europe, taking the Cape route around Africa. But many of these giant ships unload their cargoes at Mediterranean storage and refining centers and pipeline terminals.

There is also a substantial amount of oil in Mediterranean countries. During 1974, Libya and Algeria alone produced an estimated average of 2.58 million barrels a day.[5] Another 1.5 million barrels a day is pumped to eastern Mediterranean pipeline terminals.[6] All told this is about 4 million barrels a day, or about 30 percent of Western Europe's petroleum imports in 1974.[7] Even if it were possible to leave aside the powerful role of oil, it would not be difficult to illustrate how important the sea is to the economic and political health of Mediterranean countries—and from this to demonstrate why it is in the interest of the United States to maintain its vitality.

Sea Lines of Communications. Two basic facts make the sea the main line of communication for most countries of the world: water covers 70 percent of the earth's surface and it is economical to move large quantities of goods by sea. The only countries not dependent on sea communication are countries that have great continental depth and an abundance of natural resources and countries that are landlocked. But even these nations are dependent on the sea to some extent for foreign trade or as a platform for pursuing their foreign policy objectives.

In times of crisis when a massive airlift is required to quickly relieve a stranded country, the sea becomes secondary. It did, for example, when the United States resupplied Israel during the 1973 Arab-Israeli War.[8] But a sustained airlift is prohibitively costly, and

---

[5] *Oil and Gas Journal*, vol. 72, no. 52 (30 December 1974).

[6] The figure is based on averaging pipeline capacities for Iraq Petroleum Company, Tapline, and the Trans-Israeli Pipeline.

[7] A more dramatic statistic is that Italy imports an average of 2.2 million barrels a day.

[8] The United States depended heavily on the navy for the 1973 airlift both for logistics and protection. The A-4 Skyhawk jet attack fighters sent to Israel landed on aircraft carriers stationed along flight paths because of the plane's limited range and because European allies would not permit landing rights at U.S. bases on their soil. The A-4 is a U.S. Navy aircraft and one of the principal

if the sea could not be used as a route of communication it is likely that air routes would be closed too.

For most countries in the Mediterranean Basin, many of which are net importers of food and other essentials, free access to the sea is vital. The surrounding terrain is rugged—mountain or desert—and does not permit extensive road or rail links. The politics of neighboring countries are often as inhospitable as their terrain, and sometimes hostile. Even where there are overland routes, they are at the mercy of the countries through which they run. In this sense many Mediterranean states are as much islands as Malta and Cyprus.

Israel is one of the most obvious examples. Because of the Arab-Israeli conflict Israel is isolated from its neighbors. Its only open borders are the sea: the Mediterranean and the tiny outlet to the Red Sea and Indian Ocean through the port of Elath. Even though most of Israel's trade is with countries west of Suez, the bulk of its petroleum imports are delivered by tanker at Elath. Nasser's threat in 1967 to close the Strait of Tiran and cut sea traffic to Elath was a major cause of the war that followed.

Look at Lebanon. The Arab-Israeli dispute prevents any substantial traffic across its southern border with Israel.[9] Syria, to the east and north, does not have formal diplomatic relations with Lebanon. For several prolonged periods since the end of the French mandate after World War II, Syria has closed its border with Lebanon and blocked goods in transit—many of them perishables—to Jordan, Saudi Arabia, and Arab states in the Persian Gulf. Quite apart from bringing political pressure to bear on Lebanon, the Syrians were sending another message: why bother with the uncertainties of using Beirut when the Syrian ports of Banyas, Tartus, and Latakia are available?

Greece, a country in which the Western alliance has a vital stake, provides another illustration. Its economy, and therefore its political stability, is based on the Mediterranean Sea. The sea is the major conduit of its foreign trade (mostly imports) and the only certain one if road and rail routes through Turkey, Bulgaria, Yugoslavia, and Albania are cut. All of these neighboring countries have histories of antagonism toward Greece. Nearly half of Greece's 9 million people live in greater Athens-Piraeus, Salonika, and Patras—all of them

---

attack planes of the Israeli air force. Other aircraft carriers in the eastern Mediterranean provided protection for C-5As that were carrying material to Israel. There was a proposal—firmly rejected by the Egyptian high command—that Egyptian MiG-21 interceptors shoot down one of the giant transports.

[9] There is limited movement across this frontier mostly by United Nations and Red Cross personnel.

Table 2

MEDITERRANEAN POPULATION FIGURES

Country	Number
Albania	2,420,000
Algeria	16,930,000
Cyprus	660,000 [a]
Egypt	37,520,000
France	52,470,000
Greece	9,020,000
Israel	3,360,000
Italy	55,500,000
Lebanon	3,230,000
Libya	2,230,000
Malta	319,000 [a]
Morocco	17,320,000
Spain	32,610,000
Syria	7,370,000
Tunisia	5,750,000
Turkey	39,910,000
Yugoslavia	21,400,000

[a] Figures from *World Almanac 1975.*

Source: *The Military Balance 1975–76* (London: International Institute for Strategic Studies, 1975).

major Mediterranean ports. The bulk of their populations are employed directly or indirectly in trades tied to the sea: shipping, ship building and repair, fishing, banking, and insurance. Many of these people also belong to unions that are one of the most influential centers of power in Greece. And hundreds of small coastal towns and islands depend on the sea for survival. Any break in the unlimited access to the sea that Greece now enjoys would be immediately felt and could generate dangerous political pressures in a country where the domestic outlook is uncertain.

Italy is still another example of a Mediterranean country heavily dependent on the sea and one where there are apparent political dangers to Western interests. Barely anchored to the European continent, mainland Italy has 1,542 miles of Mediterranean coastline, about two-thirds of its border. Italy has 29 major ports, more than any other country in the basin. There are an additional dozen or so "specialty ports" that handle only oil or specific kinds of ore or grain.

13

No matter whether Italy continues to enjoy good relations with its neighbors—Yugoslavia, Austria, Switzerland, and France—and no matter whether its land lines of communication remain open, Italy could not rely on the sea more than it already does. Any protracted disruption of movement in and out of Italian ports would send shock waves through Italy and might further aggravate the country's political malaise. It would certainly place additional strains on Italy's political and economic system. In the best of times the Italian Communist party is influential, though it is not a member of the government, and the neofascist movement is growing. One only has to speculate what part these factions would play in times of extreme turmoil.

Algeria, Tunisia, Libya, Syria, Turkey, Albania, and Yugoslavia are countries in similar situations. Algeria, Tunisia, and Libya have the desert at their backs and the sea in front. The others either have access to the sea or become hostages to their neighbor's politics.

A few Mediterranean countries, of course, do not rely exclusively on the Mediterranean, but all would experience great political risk and economic vulnerability if the sea were unavailable. For example, although Morocco, Spain, and France also have Atlantic coastlines, important areas in all three countries would be affected by a loss of access to the Mediterranean. For Spain the large coastal cities of Malaga, Valencia, and Barcelona would suffer, and the Balearic Islands and the Spanish enclaves in Morocco would be cut off. For France, the ports of Marseille and Toulon would be crippled and Corsica isolated from the mainland. All of northern Morocco would also be affected. Egypt is another exception because it has a frontier on the Gulf of Suez and the Red Sea. But, like Israel's, the overwhelming bulk of Egypt's imports including its military wherewithal arrives at Mediterranean ports.

The Role of Naval Power

This analysis compels us to conclude that the Mediterranean Sea is indispensable to the countries it borders. Even the countries with other coastlines depend on the Mediterranean to some extent. In an environment so dominated by water, the role of naval power is of paramount importance, and rather than losing value in an age of lasers, space flight, and solid state electronics, navies have gained in prominence. There are several reasons for this development. First, because increasing attention is being directed to the sea as a potential source of food, minerals, and energy, the fields of ocean science and technology and maritime law are growing in importance—and the principal instruments for exploring, exploiting, and protecting ocean resources are ships and navies.

Second, navies make a unique contribution to warfare and to supporting nations' foreign policy objectives. In general, they provide a relatively independent, roving weapons platform that can operate long distances away from its home waters to insure freedom of the seas in times of peace and control of the seas in times of crisis and conflict. Port calls and "showing the flag" by naval vessels are accepted traditional practices in international relations. Just as customary—and universally understood—is "gunboat diplomacy," the assembling of a fleet in international waters close enough to a nation's coast to silently demonstrate political interest and intent to use force to bring about a desired result. A fleet can cruise within sight of the shore and be visible to the general population or wait over the horizon, depending on the level of pressure a nation wishes to exert. If the fleet includes an aircraft carrier or two, the show of force or threat to intervene is made that much more credible.

Finally, almost every kind of military system, from infantry to armor to aircraft to intercontinental ballistic missile, can be launched from a ship. Indeed, the foundation of both superpowers' strategic deterrent policy is the relative invulnerability of the submarine-launched ballistic missile system.

None of this is to say that a navy is the perfect or complete military organization. Navies cannot field large armies to assault and hold great swaths of territory, though armies may be transported in ships and other naval vessels may protect troop convoys. Nor do carrier-based aircraft have the long-range and heavy-weapons-loading capability of land-based bombers. But neither armies nor air forces have the ability to hover offshore or flexibility enabling them to apply several levels of nonviolent pressure. Armies do not visit, they must either invade or be stationed at permanent bases, a form of presence that is becoming less and less palatable in today's world. Neither do air forces visit, they bomb and strafe, unless they too are permanently based. Shore batteries can provide coastal defense, but the range of guns is limited. Patrol aircraft can be used for maritime surveillance, but to rely on them exclusively would be very costly, and aircraft are limited by distance and crew fatigue. The necessity of obtaining international overflight rights is another restriction on aircraft.

What makes the Mediterranean so different, one may reasonably ask, from other oceans and seas such as the South Atlantic? There are several important differences. First, trans-ocean distances are greater, preventing any real contact between Latin America and Africa. The South Atlantic is not an enclosed sea. There are sharp cultural and linguistic differences, and no history of trade and colonization unites the two shores. One shore does not depend on the mineral wealth of the other for its energy. There is not the same keen strategic competition or convergence of vital interests between the two superpowers. There are no American military bases in South America or along the west coast of Africa that contain weapons capable of reaching the Soviet Union. And the Soviet navy does not permanently maintain a fleet in the South Atlantic. The reverse circumstances exist in the Mediterranean, and it is in this context that the balance of military power within the sea ought to be considered and evaluated.

3

THE UNITED STATES PRESENCE IN THE MEDITERRANEAN

Overview and Background

The Mediterranean Basin often reflects the state of relations between the two superpowers, and the United States and the Soviet Union seem to have chosen the Mediterranean as an arena for contest: the most poignant illustration of these principles is the massive stock of weapons that the Mediterranean countries have introduced into the region—both the weapons possessed by their own armed forces and the small but powerful arsenals they have created in other countries through military assistance programs.

Another, less visible example of this competition is the intelligence-gathering activities of both superpowers as they try to ascertain each other's military capability. Some ships, aircraft, and land-based facilities, all of them sprouting with antennae and equipped with banks of electronic consoles, spend their entire time probing, ferreting, listening, and recording. The reason for this effort is, of course, to maintain or gain military advantage. Of the two superpowers, the United States has the more elaborate and sophisticated military presence.

All told the United States has 60,000 men, 275 combat aircraft, and 45 ships in the Mediterranean Basin. But these numbers, which have to be approximate because precise strengths change, are not an accurate gauge of American power. On paper the United States may seem weak compared to other countries around the sea. However, the test is what these forces and weapons can do.

For example, a U.S. Lafayette-class submarine armed with 16 Poseidon missiles (the shorthand for Polaris/Poseidon submarines is SSBN) cruising below the Mediterranean's surface is capable of hit-

17

ting targets 2,500 miles away, which puts the heartland of the Soviet Union well within its range. The addition or withdrawal of one such submarine makes a significant difference in strategic terms, but would alter the numbers only marginally—140 men, one ship, no aircraft. Even the number of missiles could be misleading because each Poseidon missile can contain up to ten nuclear warheads, each of which can be directed at a different target, giving that one ship more firepower than all the armed forces of the Mediterranean countries combined.

Sixth Fleet aircraft, flying from the two carriers normally on station in the Mediterranean, can deliver conventional and nuclear weapons more than 1,000 miles away from their bases. The addition of a third carrier increases fleet strength by just one ship but adds 90 aircraft. That happened during the Middle East crisis of October-November 1973, when the *John F. Kennedy* joined the Sixth Fleet in the eastern Mediterranean. A floating reinforced marine battalion of 2,000 men with their own artillery and armor—less than 5 percent of the 60,000-force level—can land anywhere in the Mediterranean while Sixth Fleet planes patrol overhead.

Most of the American troops based on land man a logistics network that can provide and replenish everything from butter to bullets to ballistic missiles. Others operate small communications installations and electronic surveillance facilities, some within eyesight of the Soviet Union and Bulgaria, that yield valuable intelligence on Warsaw Pact military movements and strengths and furnish early warning of Russian missile launchings.

This, of course, is only a bare summary to show that numbers are far less important than capabilities and to provide a context for understanding what follows—a systematic country-by-country survey of American forces.

The Base System

Portugal. Though not geographically a Mediterranean country, Portugal has played an important role as the first link in the chain of American bases in that sea. Lajes Air Force Base on Terceira Island in the Azores group of islands, about 1,000 miles due west of Lisbon, has been a first stop for trans-Atlantic supply flights to Europe and the Mediterranean and a base for U.S. Navy maritime (antisubmarine warfare, ASW) patrol aircraft. C-5As stopped there for refueling during the emergency airlift to Israel during the 1973 Middle East war, making Portugal the only European country to openly permit

American planes to land. As a result Portugal was subjected to a complete oil embargo by the Arab petroleum-producing countries and suffered acute fuel shortages.[1]

Since the 1973 war, however, Portugal has undergone a dramatic political change. In 1974, the army overthrew the right-wing regime and moved to the left in domestic politics, causing concern in Washington about Portugal's future as an ally and its role in NATO. In April 1975, the Portuguese government said that it would not allow Lajes field to be used by U.S. aircraft to resupply Israel during any new Middle East conflict.[2] The availability of Lajes to American forces at least in a non-NATO Mediterranean context may be somewhat moot since the C-5A has now developed an in-flight refueling capability enabling it to fly nonstop from the continental United States to the eastern Mediterranean. But the cost of doing so is higher and the cargoes must be smaller than when a stop in the Azores is possible. The United States acquired the Lajes base in 1946. The latest formal agreement expired in February 1974, but use of the base will continue until new terms are negotiated. About 1,000 American servicemen are stationed at Lajes.

Spain. Moving from west to east, the first European country with a Mediterranean coastline is Spain. It has the second largest number of American forces stationed in any country in the basin. There are 10,000 American servicemen, plus another 13,000 military dependents in Spain, and four bases, two of which are major.

The most important base in Spain, because of its role in American global strategic planning, is Rota, a sprawling U.S. Navy nuclear missile submarine service station just west of the Strait of Gibraltar.[3] Activity at Rota is built around a submarine tender—the U.S.S. *Holland* when this writer was there in 1972—moored in the harbor. This giant ship, designed specifically to service SSBNs, can perform every conceivable repair and maintenance operation on these submarines. It is literally a floating machine shop with a nuclear dimension, including facilities for storing and servicing the ballistic missiles and their warheads.

Security on the submarine tender boggles the mind and conveys a sense of how the United States views its mission and the role of the SSBNs that patrol in nearby waters. U.S. Marines armed with

---

[1] *New York Times*, 9 April 1975.

[2] Ibid.

[3] Rota is one of three such bases outside the continental United States. The other two are located at Holy Loch, Scotland, and Apra Harbor, Guam.

19

shotguns lowered at the ready stand guard over the missile storage tubes and the ship's stern boarding-access ladder. Every person on the tender must display a badge or have an escort at all times, in some cases both. All personnel must also wear visibly radiation-sensitive medallions that are checked periodically to insure that personnel have not been exposed to toxic or lethal levels of the nuclear material they work with.

There is also an airfield at Rota that is used primarily by transports and maritime surveillance aircraft, though navy and air force fighters occasionally land. And Rota has a fuel and munitions storage depot and a communications station.

The other major American installation in Spain is Torrejon Air Base, northeast of Madrid, which was the headquarters of the U.S. 16th Tactical Air Force until it moved to Naples in 1974. The 16th Air Force, most of whose planes are stationed at forward bases in Italy and Turkey, is a mixture of F-4 Phantoms and F-111 fighter bombers. Some of the units are still based in Spain. Torrejon is also used by KC-135s, the flying tanker version of the Boeing 707. These tankers used Torrejon to refuel U.S. fighter jets being sent to Israel during the 1973 Arab-Israeli war.[4]

There are two other smaller installations, the air bases at Zaragoza and Moron. Zaragoza, in northeast Spain between Madrid and Barcelona, is used by other U.S. Air Force units stationed in Europe during bombing and gunnery practice. On a rotation basis these jet fighters from the 3rd Air Force in Britain and the 17th in Germany visit Zaragoza and compete in weapons training exercises periodically. Zaragoza was reactivated in 1969 when Wheelus Air Force Base, near Tripoli, Libya, was closed after Colonel Qaddafi came to power. The other facility, Moron, near Seville in southern Spain, is rarely used and is manned only by a skeleton crew. Officially it is in mothballs.

Italy. There can be no simple characterization of the American military presence in Italy, though many different things can be said about it. Italy has a larger number of American forces on its soil than any

---

[4] Spain, because of its pro-Arab policy, refused to give permission to land to the fighter planes headed for Israel, though it did not ask the United States to suspend the operation of the KC-135 tankers. One could speculate about the attitude of a different government, one less allied with the United States, and the effect of a less cooperative Spanish government on the balance of power in the Mediterranean. The *Washington Post* reported on 23 February 1975 that Spain is asking the United States to close Torrejon in an effort to lower the American profile in Spain and defuse Spanish criticism of the American presence. This question will be explored in greater detail in a later chapter.

other Mediterranean country—12,000 men on the Italian mainland and the two islands of Sicily and Sardinia. Most of the installations are on or near the coast, that is, most are naval bases or facilities.

The reason for such a large American contingent in Italy is directly related to France's decision in March 1966 to withdraw from integrated NATO military planning. France also asked for the withdrawal of Allied military headquarters and troops, requiring the United States to find another base for these forces. Italy was the natural choice. Italy had a strong commitment to NATO; there were many existing airfields or sites that could be easily improved; it had a relatively high standard of living which would not present a major adjustment problem for the families of servicemen; and there were strong cultural links between the United States and Italy which made the average Italian psychologically receptive to the average American.

Italy's location in the middle of the Mediterranean is another factor that determines the size and type of American forces. Maritime reconnaissance and ASW aircraft based in Sicily can conveniently monitor Soviet naval activity and other military movement in both basins of the sea. Also, it is easy for naval units stationed there to move quickly to either basin for training exercises or patrols, and for repair and maintenance facilities in Italy to be easily reached from both sides.

For example, the Sixth Fleet flagship, traditionally a guided missile cruiser with the sophisticated communications gear necessary for a command ship, is homeported in Gaeta, just north of Naples. (Before the French disengagement from NATO, the flagship's home was Villefranche on France's Mediterranean coast between Nice and Monaco.) A submarine tender that services nuclear-powered hunter-killer submarines (SSN) is homeported at La Maddalena, a tiny island off Sardinia. Let us look at these facilities and others in more detail.

The hub of U.S. activity in Italy is Naples where there is a major headquarters and support complex of the Sixth Fleet and an air force command center. The navy commander with the widest responsibility is a rear admiral who directs American surface surveillance and anti-submarine warfare efforts in the Mediterranean and is the custodian of all U.S. Navy facilities in the basin. In a NATO war scenario this same officer becomes the allied commander of ASW and surveillance. In navy parlance, he "wears three hats," which means that he has three different jobs.

As commander of Task Force 67 (TF-67), he is responsible for keeping track of all shipping in the Mediterranean that is of interest to the United States. This can include monitoring the movements of

21

anything from an innocuous looking Soviet trawler that is really an electronic intelligence collection ship (AGI), to the *Moskva*, an ASW helicopter carrier, to a Bulgarian merchant ship carrying tanks to Algeria. Under his command are P-3 Orion ASW aircraft based at Sigonella, Sicily. He can also assign destroyers fitted with special electronic gear and submarines to follow ships of high interest, including Soviet submarines. This aspect of U.S. naval operations will be fully examined in the course of the discussion of the Sixth Fleet. Another of his jobs is to oversee the operation of all U.S. Navy facilities in the Mediterranean. The Fleet Support Office in Greece, for example, is under his command, and he was the principal U.S. negotiator of the homeporting agreement there.

The headquarters of TF-69, the force of nuclear-powered hunter-killer submarines (SSN) patrolling the Mediterranean, also part of the Sixth Fleet, is in Naples. And Naples is the homeport of a destroyer tender. This ship often goes to other parts of the Mediterranean, the areas where the destroyers operate.

A submarine tender for the SSNs is anchored off the coast of La Maddalena island, part of an archipelago between Sardinia and Corsica. There are also storage and other facilities in the nearby island of Santo Stefano, to the south. About 800 dependents of the ship's crew live in the area. The announcement of the stationing of the tender brought a sharp reaction in Italy from Communist and non-Communist quarters.[5] Opposition centered on the dangers of radioactive leakage from their nuclear power plants and torpedoes. The U.S. Navy assured Italy that the submarines would not come close to the shore; hence the tender is always at anchor in La Maddalena Bay.

Because Italy is a NATO member, tactical nuclear warheads for surface-to-surface missiles and artillery are stored within its borders. Such weapons are normally kept at installations where Italian and U.S. army troops share responsibility for security: Italian troops guard the outer perimeter of the base while American forces control access at the inner fence and at the storage bunkers themselves.[6]

There are two U.S. Army bases in Italy, Camp Ederle and Camp Darby. Camp Ederle, at Vincenza, 40 miles west of Venice, is the home of the 1st Airborne Battalion Combat Team/509th Infantry and the 2nd Battalion/30th Field Artillery, two highly mobile fighting units that can be deployed—by parachute if necessary—anywhere

---

[5] *New York Times*, 2 October 1972.

[6] Jeffrey Record, *U.S. Nuclear Weapons in Europe* (Washington, D. C.: The Brookings Institution, 1974), p. 29.

in the Mediterranean Basin or in Central Europe. Camp Darby, at Livorno (Leghorn), 200 miles north of Rome on Italy's west coast, is a logistics storage and command headquarters with an army communications facility and a hospital. There are also two U.S. Air Force bases, Aviano at Roveredo In Piano, northwest of Trieste in the foothills of the Alps, and Udine, that are used as advance bases for units of the 16th Air Force.

Greece. While Italy's location makes it an ideal place from which to monitor both basins, Greece's position in the eastern half of the sea makes it a suitable location for forces that could be easily deployed in the area where most of the conflicts in the Mediterranean have occurred since World War II, and where the Soviet navy is concentrated.

There are other considerations. Once Soviet naval vessels pass through the Turkish Straits, they must continue through the Aegean Sea. Greece is next to Bulgaria, a Warsaw Pact member, and is close to the Soviet Union. Salonika Air Base, for example, is only 75 miles from the Bulgarian border and 475 miles from Odessa, a major Soviet industrial center and port on the Black Sea. Thrace—that strip of land stretching from Thessalonika to Istanbul and straddling both Greek and Turkish territory—would probably be the first major military objective in the southern theater.

Many of the American installations are in the greater Athens-Piraeus area or nearby. At Athens International Airport, the U.S. Air Force operates a base facility that supports Military Airlift Command (MAC) planes and C-130 and F-4 electronic and photographic reconnaissance planes, and serves as a land link for the Sixth Fleet units in the area. In downtown Athens the navy had a support office, a headquarters that provided personnel and administrative services to the U.S. Navy in Greece. Elefsis, Greece's major shipbuilding and maintenance area near Piraeus, was the homeport of Destroyer Squadron 12 (DESRON 12), the six destroyer escorts that constitute part of the task group built around an aircraft carrier.[7]

---

[7] On 29 April 1975, Greece and the United States announced that the homeporting arrangement would be terminated by the end of 1975 when the last of the six destroyers was scheduled for reassignment. The vessels would be replaced by rotation from Norfolk as they were before homeporting came into effect. At the same time, press accounts of the end of homeporting also said that the U.S. air base at Athens International Airport would close. Nominally, there will no longer be a portion of the base under American control; however, the roles of American planes and personnel at Athens airport will continue but under the administrative umbrella of a Greek commander and from a Greek base. *New York Times* and *Washington Post*, 30 April 1975.

Outside of Athens in Nea Makri is a major communication station for the U.S. Navy located in the eastern Mediterranean. During the November 1973 bloodless coup in Athens, this station supplemented the U.S. embassy communication links with Washington because of the volume of traffic and frequency of power blackouts.

The Athens area is also the headquarters of the 558th Artillery Battalion, assigned to Greece under NATO defense agreement, though most of their sites are located in Thrace within striking distance of the Bulgarian border. The same assumption about the storage of NATO tactical nuclear weapons must be made for Greece as for Italy.

In Crete, there are two major American facilities. At Souda Bay, the U.S. Navy maintains a large NATO supply base for the Sixth Fleet. At Iraklion the U.S. Air Force has a base for reconnaissance planes and KC-135 tankers for missions in the eastern Mediterranean. Also in Crete, at Namfi, there is a NATO missile firing range used by Allied fighter aircraft to practice combat tactics.

There are also several electronic and communication monitoring sites, some on the fringes of northern Greece where the Greek border runs along the Albanian, Yugoslav, and Bulgarian frontiers. One such site operated by the U.S. Air Force is on Lefkas island in the Ionian Sea.

Mention should be made of the Voice of America broadcast and relay radio stations in Greece. Though not military installations, they are part of the overall U.S. strategic posture and ought to be cited if only because their American employees could be the object of a rescue operation by military force. These facilities are located in the northern Greek cities of Salonika and Kavalla and on the island of Rhodes off the southwest coast of Turkey.

Homeporting. Any examination of the American presence in Greece must include a discussion of homeporting—the subject of intense controversy in both Greece and the United States. At times the debate has been so shrill in Washington that it has baffled many Greeks and caused them, even those not enthusiastic about the proposal, to question the strength and resolve of U.S. power.

Homeporting in Greece is a complex issue. The U.S. defense budget, American policy toward Greece (or, rather, the glaring uncertainty about what that policy ought to be), unsettled Greek domestic politics, and U.S. Navy reenlistment and training requirements bear on this question as much as the strategic arguments for having an aircraft carrier and a destroyer squadron stationed in the eastern Mediterranean. Each of these factors was involved in the unsuccessful efforts to conclude a full homeporting agreement with Greece.

Because of budget pressures, the Defense Department decided to reduce its attack carrier force from 15 to 12 ships.[8] The navy uses a patrol-to-transit/repair ratio of one to two; that is, for every carrier on patrol, two others are either sailing to or from their operational area or are out of service for maintenance or training. These figures apply to normal peacetime operations, with three carriers on assignment in the Pacific and two in the Mediterranean. In times of crisis or war this formula could change.[9]

Rotation of the ships means that the crews would have to be away from their families for prolonged periods, usually six to nine months. With a reduced carrier force the periods of separation could be extended, with serious implications for reenlistment rates. One of the reasons men consider leaving the navy—apart from a very positive incentive, that their skills are in high demand in civilian life—is the clear disincentive of long absences from their families.

The navy requires personnel with relatively high levels of skills to operate the sonar, radar, and other complex electronic equipment aboard its ships. It takes from two and a half to three years for a man to complete basic training, finish technical courses, which usually include advanced training, and get the necessary on-the-job experience to make him a fully competent sailor in today's navy. If the man decides to leave at the end of his four-year enlistment, the navy gets only one full year's use of his skills at peak performance.

Faced with this situation, Admiral Elmo R. Zumwalt, Jr., the highly innovative chief of naval operations between 1970 and 1974, made his proposal for homeporting. The original proposal from the navy was to station a carrier and six destroyers in Greece for two years and move the families of the men (about 4,000 dependents) to Athens, a metropolitan area with a population of 2.5 million. All told, however, this would have placed an additional 11,000 servicemen and dependents in Greece, bringing the total U.S. military presence there to 16,000. The navy would have required a minimum of shore-based facilities for these ships; a pier and a few storage buildings, not much more. A low profile could have been achieved

[8] These budget pressures were so severe that this factor, coupled with the age of many ships, reduced the size of the navy from 934 vessels in January 1969 to 594 in July 1973.

[9] For a fuller discussion of the impact of reducing the attack carrier forces, see Norman Polmar, "New Carrier Concepts and CVN-70," in *Sea Power*, vol. 15, no. 3 (March 1972).

by having the ships dock at Elefsis near the Greek naval base—more or less out of sight of the public at large.

The plan was divided into Phase I and Phase II. The first phase included stationing the six destroyers at Elefsis and moving the crews' families to Athens. Phase I also included moving the families of the staff of the carrier task force commander. This was implemented.

The second phase would have involved basing the carrier. The plan also called for stationing a family service ship to provide shopping and medical clinics for the civilians so that the PX and commissary facilities of other American military units in the area would not be overburdened. Phase II, which was the more controversial, was never implemented even though alternative sites, such as Souda Bay, Crete, were considered.

Right away the proposal ran into opposition. First there were those in Washington and Athens who felt that any increase of American forces in Greece would be a clear signal that the United States wanted the military-based regime to stay in power and would give credibility to leftist allegations that the United States helped bring it to power in the first place. Further, critics said, any homeporting agreement concluded at that time (1971–74) would not encourage the return to parliamentary democracy in Greece, which was the announced policy of the U.S. government.

Though the homeporting plan was official U.S. policy—the navy had not entered into negotiations with Greece on its own—the proposal never had ardent support in the State Department. Its officials gave the proposal only passive approval, largely because they were not prepared to get into a fight with the navy over homeporting.

In a wider context, the State Department itself was divided over policy toward Greece. The dominant view was that the United States ought to be pragmatic about the presence of the military government in Athens and continue to do business with Greece in a cordial atmosphere. This way the United States would avoid damage to vital bilateral and Allied security interests while it quietly encouraged a return to elective government. Others in the State Department felt that the tone of Greek-American relations ought to be just barely civil—no arms sales, no homeporting. This split was often apparent in Athens, where some of the U.S. embassy staff held the latter view.

An episode illustrating these differences occurred in 1974 when the embassy advised the commander of the Sixth Fleet not to attend a ceremony marking a Greek national holiday. However, another American admiral, the commander of NATO's southern forces based in Naples, arrived on the scene, leaving a chagrined embassy staff with a fait accompli. As an officer in the NATO chain of command, he effectively circumvented the diplomats.

There was also resistance to the plan within the navy because the Sixth Fleet is represented as a force not needing facilities ashore to carry out its mission. That is only partially true. The Sixth Fleet may not need shore facilities, but it has always used land-based installations—fuel depots, electronic and communication sites, and air bases. The homeporting plan ought to be seen also as an attempt to secure facilities on the northeastern littoral after the U.S. Navy lost its access to the southeastern Mediterranean to the Soviet navy. A minor theme of the resistance was opposition to Zumwalt himself. Homeporting was just one in a series of policies that drew fire from his critics.

Inside Greece resistance to homeporting took many forms. Because the homeporting talks straddled the oil crisis following the 1973 Arab-Israeli war, the Greek government became more sensitive to Arab charges that the United States was building another "base" in the Mediterranean and on Greek soil. Also, after Brigadier General Dimitrios Ioannides replaced George Papadopoulos as Greek strongman in November 1973, the government indicated that the terms for concluding the second phase of homeporting must include more military hardware for the Greek armed forces.

The opposition press—forbidden by stringent censorship from criticizing the government—bitterly attacked the homeporting proposal, though not directly. Rather, their targets were the American sailors who got into fights with taxi drivers or who were arrested for drunk and disorderly conduct. Ironically, most of these sailors were not "homeporters" but were from the crews of American navy ships making casual port visits to Greece.[10] Inflation in Greece was rampant and the press editorialized that a major cause of the soaring prices was the influx of navy families. The press did not report that at the time the number of additional American families was less than 500. In short, the U.S. Navy and homeporting became surrogates for the military regime the press did not dare criticize.

The future of the U.S. presence in Greece now hinges on the course of events in Cyprus. Homeporting is the first casualty. Indeed the future of the American presence in the entire eastern Mediter-

---

[10] The *New York Times*, for example, reported on 26 December 1972 that the Greek press gave wide coverage to the charge of assault brought against an American sailor for punching a taxi driver in a fare dispute and presented the incident as "an offense to Greek tranquility." But similar treatment was not given to a Christmas party for children at a Greek orphanage organized by the crew of the U.S.S. *Sampson*, one of the destroyers homeported at Elefsis.

ranean is linked to what happens in Cyprus because the bases in Turkey have been jeopardized; the U.S. Congress voted to cut off military assistance to Turkey because no progress was being made in settling the dispute.*

When the military government stepped aside in July 1974, leftist opinion in Greece began to call openly for the withdrawal of American bases, largely as a reaction against the previous regime which, it was widely believed, the United States had supported. The call by the new government for a review of the American presence was part of this reaction. But a deeper, more substantive debate is also going on about the future of Greece and its close ties with the United States and NATO. Not all of the anti-American opinion is leftist; in fact, much is found in highly nationalist and rightist quarters, for example, among young army officers. At mid-1975, the feeling was that the more important U.S. bases—those that contribute to Greek as well as NATO security—would remain. Other casualties could be the facilities that the United States uses to support its Middle East policies—like the base on Crete from which the KC-135 planes staged their midflight refueling missions to planes carrying supplies to Israel.

Turkey. The composition of the American military presence in Turkey is largely determined by Turkey's strategic geographic location. It is next to the Soviet Union. The Turkish Straits are the major route used by Soviet navy ships (except submarines [11]) to reach the Mediterranean and to ship arms and other military equipment to the Middle East and North Africa. With the Suez Canal open the Turkish Straits-Mediterranean route is the shortest way for the Soviet navy to reach the Indian Ocean. One American scenario for war with the Soviet Union assumes that Russian and Bulgarian forces would attempt to move across Greece and Turkish Thrace to obtain direct and unimpeded access to the Mediterranean.

These facts and assumptions govern a very special relationship between the United States and Turkey and determine the nature of military and security assistance—in both directions. Turkey, of course, is a member of NATO.[12]

---

* Editor's Note: This section was written before Congress voted to partially lift the arms embargo against Turkey.

[11] Black Sea powers are prohibited from using the Turkish Straits to deploy submarines in the Mediterranean. See Article 12, Montreux Convention, in appendix.

[12] For a discussion of American military presence in Turkey in the context of the straits, see Ferenc A. Vali, *The Turkish Straits and NATO* (Stanford, Cal.: Hoover Institution Press, 1972).

Perhaps the most substantive measure of Turkey's importance to the United States and its role in U.S. defense policy is the fact that from 1946 to 1973 Turkey received $3.72 billion in U.S. military aid. For the years 1962 to 1973 the amount of U.S. aid was $1.857 billion—more than any other country in Europe received in that period.[13] The number of American forces in Turkey is relatively modest—7,000—when compared with the U.S. presence in other Mediterranean countries.[14] Today the United States has troops at more than 20 bases in Turkey, most of them stationed at facilities called "common defense installations" under joint U.S.-Turkish control. In addition, the U.S. navy often takes on fuel from a NATO naval bunkering site near Iskanderun, just above the Syrian border.

The principal installations in Turkey are electronic early warning and monitoring sites and air bases. These are used by forward units of the 16th Air Force, reconnaissance aircraft and by the United States Logistics Group (TUSLOG), the American transport and supply network that serves Turkey and the Middle East. Among these facilities are:

(1) Sinop on the Black Sea coast almost directly opposite Yalta. This is a vital electronic monitoring station.

---

[13] Agency for International Development, Statistics and Reports Division, Office of Financial Management, *U.S. Overseas Loans and Grants and Assistance from International Organizations, Obligations and Loan Authorizations, July 1, 1945-June 30, 1973,* May 1974, p. 28.

[14] This is a sharp reduction from previous force levels, largely due to across-the-board cutbacks of overseas deployment and Turkey's growing sensitivity to a large, conspicuous American presence. In 1963, for example, the number of American military personnel and civilian employees of the Department of Defense stationed in Turkey was 12,275. See U.S. Congress, Subcommittee on Arms Control, International Law and Organization, Senate Committee on Foreign Relations, *Forces in Europe: Hearings,* 93rd Congress, 1st session, 25 and 27 July 1973, pp. 163 and 227. Also see Roland A. Paul, *American Military Commitments Abroad* (New Brunswick: Rutgers University Press, 1973). Paul states that in 1967-68, there were about 24,000 American troops and dependents in Turkey (p. 171). Roughly half of them were military personnel. Until early 1963, the United States had Jupiter intermediate-range ballistic missiles (IRBM) with nuclear warheads based in Turkey. They were withdrawn as part of a settlement of the 1962 Cuban missile crisis, though their withdrawal was something less than a quid pro quo for the Soviet Union. By early 1963, the first Polaris submarines were being deployed in the Mediterranean. They are a far better deterrent than the Jupiter rockets. The precise locations of the submarines are virtually impossible to determine. The Polaris missile burns solid fuel, can be launched instantaneously, and has a greater range. The Jupiter, by contrast, was fired from known, fixed sites, required time to pump in its liquid fuel, and had a relatively short range. The amount of time it took to prepare the Jupiter for launching made it vulnerable to a preemptive strike or sabotage.

(2) Diyarbakir, an electronic monitoring post just west of Lake Van in southeast Turkey near the Syrian border. This site is near Pirincilik Air Base, another U.S. facility.

(3) Incerlik Air Base, located outside the city of Adana in south central Turkey and very close to the Syrian-Turkish frontier.

(4) Karamursel Air Base, near the town of Izmit southeast of Istanbul and at the head of a long, narrow bay of the Sea of Marmara. This base could be used to guard against any threat to the straits. A U.S. Air Force communication facility is also located here.

(5) Cigli Air Base, near Izmir on Turkey's Aegean coast. Izmir is also the location of the 6th Allied Tactical Air Force, essentially an American unit composed of 16th Air Force planes that could be assigned to NATO in case of war.

Some of these bases have played a variety of roles in executing U.S. national security policy. In 1958, during the landing of U.S. Marines in Lebanon, air force planes brought 1,600 U.S. Army troops from Germany to Incerlik to have as a ready reserve for possible use in the Arab world.[15] Reconnaissance aircraft have also used bases in Turkey. For example, the ill-fated flight of Gary Powers, whose U-2 was shot down over the Soviet Union on 1 May 1960 while on a photographic mission, started at Incerlik, the base of a U-2 squadron.[16]

The most visible face of the American presence in Turkey is TUSLOG, the United States Logistics Group, whose major centers are Ankara, the capital, and Izmir, on the Aegean Sea. It is operated by the U.S. Air Force. Some 24 TUSLOG operations are based in downtown Ankara, in the suburb of Balgat, and at Ankara airport, where a portion of the field is reserved for air force use. At Izmir there are warehouses, motor pools, personnel offices, schools, and medical facilities. The Izmir installation also serves the NATO LANDSOUTHEAST headquarters and the 6th Allied Tactical Air Force.[17] TUSLOG also provides refueling and replenishment services for other military aircraft flying supplies and men to Lebanon, Jordan, Israel, Saudi Arabia, and Iran.

The electronic intelligence (Elint) and communication sites operated by American military personnel are scattered in the remote and rugged terrain on the Black Sea coast and in northeastern Turkey close to the Soviet Union, close enough to listen to Soviet military radio traffic. Anything of immediate tactical value learned from this electronic eavesdropping can be passed quickly into the American

---

[15] *New York Times*, 20 July 1959.
[16] See *New York Times*, 6 May 1960.
[17] Paul, *American Military Commitments Abroad*, p. 171.

and Allied defense system. But because most of the Soviet transmissions are in codes, they are recorded and sent to the National Security Agency for more thorough analysis.

Often these ground sites work in cooperation with aircraft that fly close to the borders of the Soviet Union so that Russian technicians will turn on their antiaircraft radars. The electronic transmissions of these radars are then recorded and an analysis is made of their characteristics. The results of such intelligence operations, which are conducted by both superpowers in the Mediterranean and throughout the world, can be used in many valuable ways. The American Shrike missile, for example, used to destroy the radars that guide surface-to-air missiles (SAM), was designed from information gathered by such Elint operations in Vietnam and other locations.[18]

In writing about the value of electronic intelligence, authors David Wise and Thomas Ross said that the system is

> so effective that no Soviet rocket could get off the ground without its being known within a few minutes at the North American Air Defense Command in Colorado Springs and at the CIA and the White House. The first point of detection was a radar and communications system in the Middle East. It was centered in Turkey at small Black Sea towns such as Zonguldak, Sinop and Samsun. There, powerful radars and listening gear monitored the countdowns and rocket launchings at the main Soviet missile sites near the Aral Sea.[19]

Another fascinating dimension of the American presence in Turkey is the monitoring of Soviet shipping—both naval and merchant—that passes through the Turkish Straits. The geography of the straits makes this an ideal site for gathering naval intelligence which, when combined with other data such as satellite photographs, produces highly valuable information. The straits, which are the only outlet from the Black Sea to the Mediterranean, consist of the Bosporus, the Sea of Marmara, and the Dardanelles. The Bosporus is 16 miles long and has an average width of 1 mile, though it narrows in places to less than 800 yards. The Sea of Marmara is about 130 miles long from the Bosporus to the Dardanelles and 40 miles across

---

18 The United States has supplied Shrike missiles to Israel whose pilots used them against Egypt in the 1973 war.

19 David Wise and Thomas B. Ross, *The Invisible Government* (New York: Random House, 1964), pp. 309-10.

at its widest point. The Dardanelles is 25 miles long and narrows to about 4.5 miles in the south to about 2.5 miles in the north.

Merchant shipping of any country can freely use the straits in time of peace and can pass during the night or day. Naval vessels of Black Sea countries, however, must notify Turkey at least 8 days in advance (non-Black Sea powers are asked to give at least 15 days notice) and can only transit the straits in daylight; that is, they must enter the straits in daylight and must complete their voyage before dusk the same day. This means that all Soviet naval ships and many merchant vessels can be photographed with relative ease.

Because the Soviet Union often changes the hull numbers on its warships, American analysts who monitor naval traffic in the straits have dubbed their craft "dentology" and "crateology": dents in the hull and the size and shapes of crates carried on deck are among the indicators they use to identify a ship and its cargo. These analysts can also judge the capabilities of warships by studying their arrays of antennae, weapons launchers, and guns. Because the waters are calm in the narrows, it is also possible to tell how deep a vessel is riding. This helps to determine how much cargo it is carrying. When it comes to merchant ships, the size and the shape of crates and the forms that tarpaulins take when draped over equipment also help determine the cargo. Analysts using these methods are able to monitor shipments of military hardware to Egypt and Syria.

Cyprus. There are two tiers to the American presence in Cyprus—the highly visible and the discreet. The visible and public American presence is in limbo because of the 1974 Cyprus crisis. The American civilians who operated the Foreign Broadcast Information Service station that monitors radio transmissions in the eastern Mediterranean and Middle East were evacuated with other foreign nationals living on the island in July 1974. Whether they will return and the station will reopen depends on the nature of the settlement of the three-way crisis between Cyprus, Greece, and Turkey. Until the spring of 1974 when it closed, the U.S. Navy operated a small communications station on the outskirts of Nicosia, the capital. It was staffed with about 200 men. The most interesting aspect of the American presence on Cyprus, however—and one that makes a more significant contribution to American interests in the eastern Mediterranean—is the operation that uses the British sovereign bases of Akrotiri and Dhekelia. After intense negotiations with Britain and Cyprus, the United States based high-flying U-2 reconnaissance planes at Akrotiri on Cyprus's southeast coast in 1970. One of the conditions of the agreement was that it not be made public. This operation, called "Even Steven" within the U.S. government, monitored the 1970 Suez Canal cease-fire agreement that was arranged by the United States between Israel and Egypt. The flights were tacitly approved by both Israel and Egypt, both of which felt they were necessary to preserve the cease-fire. The Soviet Union did not object to the use of Akrotiri for this limited purpose, in part because it was assured that the two parties to the cease-fire approved. However, the chance detention by Cypriot authorities of four U.S. Air Force personnel in February 1975 led to public disclosure of the flights.[20]

American servicemen, mostly navy personnel, were based in Cyprus in 1974 while helping other navy units clear the Suez Canal. U.S. Marines from a helicopter carrier needed for the Suez operation were also temporarily bivouacked in Cyprus. The marines had a relatively low profile and their presence was permitted because it facilitated work on a project of obvious benefit to all countries in the eastern Mediterranean.

Morocco. There are three U.S. Navy installations in Morocco, an air facility at Kenitra on the coast north of Rabat and two communications bases at Sidi Yahia and Bouknadel, both within a 25-mile radius of Kenitra. The main business at Sidi Yahia and Bouknadel is to provide a trans-Atlantic telecommunications link for the Sixth Fleet and for the SSBN fleet patrolling the eastern and mid-Atlantic and the Mediterranean. The Kenitra base serves the communications stations and also provides training for the Moroccan air force.[21]

The Sixth Fleet

During the discussion so far there have been frequent references to the U.S. Sixth Fleet, largely with regard to its shore-based facilities. But what about the fleet itself? What is its role in the Mediterranean? How is it organized? What are its capabilities and strengths? What are its limitations and weaknesses?

Briefly, the Sixth Fleet is both the symbol and the substance of the United States's military presence in the Mediterranean Basin. It is a highly mobile floating base that projects American power throughout the basin and beyond. Though it uses shore-based facilities, the fleet can operate for prolonged periods independent of the Mediter-

---

20 *Washington Post*, 28 February 1975. See news dispatch titled "Cyprus Seen Base for U.S. Spy Planes" by the Associated Press.
21 See also Paul, *American Military Commitments Abroad*, pp. 192-93.

ranean littoral, perhaps for long enough to restore any military or political imbalance which might deny it access to the resources of the shore.

Its two aircraft carriers convey the fleet's potent political and psychological message, its intention and ability to use its power. The carrier-based aircraft can drop conventional or nuclear weapons on any target in the sea and more than 1,000 miles from the carriers. That brings areas in the southern Soviet Union, Bulgaria, Hungary, and Rumania within reach.

Its amphibious task force can land a fully equipped U.S. Marine battalion complete with artillery and armor on any shore and as far inland as some 100 miles. This capability was the heart of a contingency plan for marines to secure an airfield for the evacuation of American citizens from Jordan if that had proved necessary in 1970. Or, as in Lebanon in 1958, the marine presence could be used to end a civil war that threatened a friendly government and to signal that the United States was prepared to use force to stabilize a volatile Middle East.

Though they do not take their orders from the Sixth Fleet commander, the Polaris/Poseidon submarines cruising the depths of the Mediterranean can deliver nuclear warheads on all of the major industrial and population centers of European Russia and many in Soviet Asia as well. That includes Moscow, Leningrad, and the Soviet strategic missile complexes.

The Sixth Fleet has the broad task of protecting American interests in the Mediterranean.[22] Under that rather large umbrella a wide variety of specific operations have been conducted, including the evacuation of Americans from Israel and Egypt in 1956, the marine landings in Lebanon in 1958, the dramatic show of force during the 1970 and 1973 crises in the Middle East, and the rescue of Tunisian flood victims in March-April 1973. The regular activities of the Sixth Fleet—engaging in joint training exercises with navies of friendly countries, honing the skills of its crews in ASW, surface, and air combat, keeping track of Soviet warships and seeking to

---

[22] According to an official U.S. Navy publication, the fleet is charged with four basic missions: "To deter aggression against Western Europe by maintaining striking forces capable of utilizing conventional and nuclear weapons and to be prepared to conduct such offensive operations as either a national or a NATO force should deterrence fail; To promote peace and stability by its readiness and availability for deployment at trouble spots; To create goodwill for the U.S. and enhance its prestige with the countries bordering the Mediterranean; [and] To protect U.S. citizens, shipping and interests in the Mediterranean area." Public Affairs Office, Staff, Commander Sixth Fleet, "The United States Sixth Fleet," undated brochure.

discover their capabilities, and making periodic port calls to "show the flag"—are all carried out with the same general objective.

Historical Background. For a full understanding of the Sixth Fleet, let us briefly look at it in a broad historical context. The United States has had naval forces in the Mediterranean Sea since the early 1800s. One of the first American naval engagements in foreign waters occurred off the North African coast when the Barbary Sultan of Tripoli—the Tripoli of the U.S. Marine anthem—declared war on the United States on 10 June 1801 because American merchant vessels had refused to continue to pay cash tributes to raiding pirates.

Between the world wars, American naval units helped settle the unstable situation in the Balkans and the Middle East. At the start of World War II, the United States was reluctant to get too deeply involved in the Mediterranean, viewing an assault on Germany from the north and, of course, the conflict with Japan in the Pacific as more urgent. After much prodding from the British, who desperately needed help to protect their own interests in the sea and east of Suez, the United States entered the Mediterranean. This was partly to relieve the pressure on the British in North Africa (Operation Torch, for example, was the Allied occupation of Morocco, Algeria, and Tunisia). But American leaders finally came to view the Mediterranean as another approach for the campaign against Germany in Europe.

The signal date for the U.S. presence in the Mediterranean as it exists today—and for the rationale behind it that is still in force—is 5 April 1946, when the battleship *Missouri* called at Istanbul to demonstrate support for Turkey against Soviet efforts to gain control over the Turkish Straits. The ostensible purpose for the *Missouri's* visit was to deliver the remains of a former Turkish ambassador to the United States, Munir Ertegun, who had died in Washington in November 1944. That port call set the stage for the proclamation of the Truman Doctrine on 12 March 1947 and other major policy statements aimed at preventing Soviet penetration of the eastern Mediterranean. In addition to putting pressure on Turkey, the Soviet Union was seeking to annex the Dodecanese Islands in the Aegean Sea from Greece, supplying Communist guerrillas fighting in the Greek civil war, supporting Yugoslavia's bid to gain control over Trieste, and refusing to withdraw Soviet troops from Azerbaijan in Iran. By early 1946, the cold war was beginning and Washington could observe a broadside political and military effort by the Soviet Union to move into the Mediterranean and Middle East. The United States chose to exhibit its resolve to check this Russian maneuver by sending the *Missouri* to Istanbul.

In June 1948, the designation "U.S. Naval Forces, Mediterranean" was changed to "Sixth Task Fleet," and on 12 February 1950, about six months after the North Atlantic Treaty went into force, the Sixth Fleet gained its present name.[23]

Fleet Organization. Now let us look at the fleet's organization, the number and types of ships and aircraft, and their individual capabilities, building a foundation of specific data before evaluating the fleet as a whole.

Carrier task force. The Sixth Fleet normally consists of 40 to 45 vessels—about half are warships—that are broken down into seven functional subdivisions, or task forces.[24] The main striking arm of the Sixth Fleet is TF-60, the two aircraft carriers and their escorts that are normally on patrol in the Mediterranean.

Each carrier has about 90 planes, about one-half of which are used for the carrier's own protection. The balance are used for strike or attack operations, though not all of these planes would actually carry weapons. Some fly combat air patrol "cap" missions—to ward off planes attempting to engage attack aircraft. Some are electronic warfare planes, which confuse ground radar, missiles, and other antiaircraft weapons. Other planes can be sent in to attack a radar site moments before the raiding planes—the ones with the bombs or rockets—make the actual strike. Still others race over the target area after the attack to take photographs so that analysts can assess the damage and judge whether another strike is required. Depending on the distances involved, carrier-based tankers, KA-3s or KA-6s, may refuel—or "hit"—aircraft on the way to or returning from their missions.

During such an operation a carrier could have 60 or 70 aircraft in the air at any one time. When on patrol in tense areas, carriers usually keep an early-warning plane and several jet fighters in the air at all times ready to react to any threat. Two other fighters, with their crews sitting in their cockpits, are always on alert near the launching catapults.

---

[23] Undated publication of U.S. Sixth Fleet Public Affairs Office.

[24] A task force is made up of two or more task groups. A carrier group, for example, consists of one carrier and between four and six escorts. These task forces are designated as TF for "task force," followed by a number. In the Sixth Fleet each task force has a numerical designation beginning with "6." A Seventh Fleet task force might be TF-77, for example.

Each carrier sails with escort vessels—frigates and destroyers armed with SAMs, powerful air and sea search radars, sonar, and other antisubmarine warfare equipment—that serve as warning and protective screens. Often one or two submarines cruise with a carrier group as an underwater escort.

The carriers themselves are a marvel of modern naval technology and are capable of operating for long periods independent of land. They are equipped with all of the necessities and amenities for a complement of 5,000 men, including hospitals, bakeries, and closed-circuit television. Precise capabilities cannot be specified, partly because such information is classified and partly because any 2 of 12 carriers of four different designs could be on duty in the Mediterranean at any given time. But it is possible to outline some of their general characteristics.

The carriers are the largest warships in the world: about 1,000 feet long, 130 feet wide, they displace 60,000 to 90,000 tons—a volume measurement. (By comparison a typical destroyer displaces only 4,500 tons.) Carriers cruise at a steady 18 to 22 knots and can reach 35 knots, even higher in bursts, though such speeds are uneconomical. Their range is more than 8,000 miles unrefueled, though all fleet operations include tankers or oilers that refuel carriers and other ships while underway.[25]

Amphibious task forces. The next two task forces are TF-61 and TF-62, which give the Sixth Fleet its amphibious and assault landing ability. TF-61 is built around various attack transports and landing craft, and TF-62 is the marine battalion combat team and its equipment. Again, at a given time any combination of ships of several designs could make up this unit.

Nearly always, however, TF-61 includes a helicopter carrier of Iwo Jima class, which are called amphibious assault ships (LPH). These ships are 592 feet long, 84 feet wide, have 18,300-ton displacement, and can travel at a steady 20 knots. They carry 2,090 troops plus 528 ships' crew. They also carry about 28 troop-carrying and 4 observation helicopters. The usual complement is 24 CH-46 helicopters that can carry 25 fully armed marines 198 miles at 143 knots and 4 CH-53 helicopters that can carry 38 marines 223 miles at

[25] Nuclear-powered ships such as the carrier *Enterprise* have an almost unlimited range. For example, in 1964 the *Enterprise*, the cruiser *Long Beach*, and the frigate *Bainbridge*—all of them nuclear powered—circumnavigated the world, cruising more than 30,000 miles, without refueling (See Capt. John Moore, RN (Ret.), ed., *Jane's Fighting Ships 1974-75* [New York: Franklin Watts, Inc., 1974], p. 408.) The new Nimitz class of carriers and other nuclear-powered vessels will have an even greater endurance.

Table 3

CHARACTERISTICS OF MAJOR U.S. CARRIER-BASED AIRCRAFT

Type	Range (nautical miles)	Characteristics
A-4	1,785 (ferry)	The A-4 Skyhawk is a nuclear-capable attack bomber that can carry up to 10,000 pounds of assorted weapons: bombs, rockets, sidewinder AAM, Bullpup ASM, gun pods, torpedoes, ECM equipment. It is a subsonic plane that can be refueled in flight.
A-6	1,880 (combat)	A carrier-borne subsonic low-level attack bomber for delivery of nuclear or conventional weapons on targets totally obscured by darkness or weather conditions. Ordnance capacity: 18,000 pounds which may include 30 500-pound bombs or 2 Bullpup missiles and 3 2,000-pound bombs. The A-6 employs a Digital Integrated Attack Navigation system plus an integrated radar display system for the pilot to "see" obscured targets. Speed: 620 mph.
A-7	2,494 (ferry)	A subsonic, single-seat, tactical fighter designed to carry a greater load of non-nuclear weapons than the A-4E Skyhawk. Its primary navigation/weapon delivery system includes forward-looking radar and air-to-ground ranging. The A-7 also contains ECM equipment: internal homing and warning systems and external pod-mounted systems. Ordnance capability: 15,000 pounds. Speed: 698 mph.
F-4 Interceptor Ground attack	1,997 (ferry) 781+ (combat radius) 868+ (combat radius)	The F-4 is a supersonic, long-range all-weather attack fighter that has highly sophisticated electronic equipment, computers and radar. It has been considered the best all-around American aircraft that can fill ground attack and all-around air-to-air combat roles but that position may be challenged by the new F-14s and 15s. It can carry nuclear weapons.
F-8	521 (combat radius)	A carrier-based single-seat fighter in service since 1957. Early models have a maximum speed of 868 knots, while late models reach nearly Mach 2. The F-8 is armed with a 4 x 20mm. Colt cannon, sidewinder missiles and other possible combinations including Matra R-530 AAMs 2,000-pound bombs, Bullpup ASMs or Zuni rockets.

Aircraft	Range	Description
F-14 (F-14A, carrier-based)	2,000 (maximum)	The latest U.S. Navy two-seat, multi-role fighter. The F-14 can be used for several purposes: as a fighter sweep/escort to clear contested airspace of enemy aircraft; as a defender of carrier task forces through combat air patrol and deck-launched intercept operations; and for secondary attack of tactical land-based targets. With a speed of over Mach 2 the F-14 has great agility in close air-to-air combat and can carry various combinations of missiles and conventional and nuclear bombs up to 14,500 pounds.
A-3 (EA-3B)	2,520 (normal)	A carrier-based electronic countermeasures aircraft with a maximum speed of 530 knots. The EA-3B has a crew compartment in its weapons bay for ECM operations. There is also a tanker version, the KA-3, for in-flight refueling missions.
E-2 Hawkeye	1,394 (ferry)	An early-warning aircraft designed for aircraft carrier use but capable of land-based operation as well. Equipped with highly sophisticated early-warning command electronics and radar systems, teams of E-2s can defend naval task forces in all weather and can be used to direct aircraft to land targets.
RA-5C	2,000 (normal)	A carrier-based tactical reconnaissance aircraft with radar-equipped inertial navigation system and bombing and navigation system. RA-5s carry reconnaissance equipment such as frame and panoramic cameras, infrared mapping and passive ECM systems. Its armament can consist of a variety of weapons including thermo-nuclear bombs. Speed: approximately Mach 2.
S-3A	2,000+ (combat) 3,000+ (ferry)	A twin-turbofan jet carrier-based ASW aircraft, more advanced in some respects than the P-3. It carries advanced sonobuoys for sensing the newer, quieter submarines and cathode ray tubes for monitoring acoustic sensors. Its magnetic anomaly detection (MAD) equipment is also more advanced for detection of submarines capable of reaching greater depths. The S-3A can be modified to perform various tasks. Plans for such modifications foresee roles as tanker, utility transport, ASW command and control and for ECM purposes. Armament includes various combinations of destructors, torpedoes, bombs, depth bombs, mines, etc. Speed: 403 mph.

Source: *Jane's All the World's Aircraft*, 1966–67 through 1973–74.

150 knots. This means that in one sortie 752 combat troops can be put ashore, say, over a distance of 50 miles in a little more than 20 minutes, while the ships cruise out of range of land-based artillery.

Other ships in this task force could be tank-landing ships (LST) of the Newport class that carry 379 troops plus their tanks, armored personnel carriers, and artillery in addition to its own crew of 213. Their cruising speed is 20 knots. Each of these ships is fitted with a huge boom and ramp that enable the vehicles and men to disembark in shallow water. The Austin-class amphibious transport docks (LPD) can carry 930 troops in addition to its own crew of 490. These ships unload by flooding the stern section and sending the men ashore in smaller landing craft carried on board. Also included would be amphibious cargo ships (LKA) with heavy-duty cranes that carry landing craft and supplies. All of these ships have large helicopter decks and carry their own helicopters.

Newer amphibious assault ships of the Tarawa class (LHA), now being constructed, are designed to combine all the functions of the above vessels. Each will carry 1,825 troops, 30 heavy transport helicopters, landing craft, tanks, other armor, artillery, and supplies, and a large hospital with surgical suites for treating wounded men. They may also carry vertical/short takeoff and landing (V/STOL) aircraft for ground support. These ships are 778 feet long, 106 feet wide, displace 39,300 tons and can cruise at 22 knots.

Supply force. Task Force 63 is composed of oilers (tankers), supply, and repair ships. They can perform most of their tasks while moving by rigging flexible pipelines for fuel and high wires for cargo nets between themselves and the recipient vessels. Helicopters shuttle between the ships to supplement this supply method. This capability, called "underway replenishment" or "unrep" in navy jargon, which brings ships as close as 80 feet while steaming normally at 12 knots, gives the U.S. Navy the mobility and independence necessary for operating for long periods in unfriendly or hostile waters. It is also a capability that the Soviet navy has not yet mastered. The tenders, however, like the destroyer tender homeported in Naples or the submarine tender located in La Maddalena, need to anchor in relatively calm waters to service other ships. But they do not require a land link; a sheltered bay will do.

Polaris/Poseidon submarines. Task Force 64 is the Polaris/Poseidon submarine force in the Mediterranean. Though these submarines are assigned to the Sixth Fleet for operational control—that is, their movements are known to a select few on the admiral's staff—the fleet commander does not give them their firing orders. Their targets and patrolling tactics are determined by strategic warfare planners through the Joint Chiefs of Staff. Sometimes a Sixth Fleet surface unit may secretly rendezvous with one of the SSBNs to take off a crew member who is seriously ill or perform some other urgent mission.

The largest single class of SSBN, which includes 31 vessels, is the Lafayette class, which displace 8,250 tons when submerged, are 425 feet by 33 feet, and carry a crew of 147. *Jane's* lists a speed of 30 knots for these ships, but they can probably go faster. They have 16 missile tubes that can carry the Polaris A-3 or Poseidon C-3 ballistic missile. Both missiles have a 2,500-mile range. All of the Lafayette-class SSBNs are scheduled for conversion to the Poseidon missile, which is fitted with MIRV warheads carrying 10 vehicles each. Each warhead usually carries several dummy or decoy vehicles to confuse antimissile defenses.

Surveillance. Task Force 67, whose commander sits in Naples, consists of maritime reconnaissance aircraft, destroyer escorts on occasion, and patrol gunboats. These units often spend days—sometimes longer—following Soviet warships and other vessels on "tattletale" and ASW missions. The destroyer escorts, for example, can monitor and record every electronic transmission of a Soviet warship.

From the data that they collect, analysts in the fleet, at the National Security Agency, and at the Office of Naval Intelligence can

---

Table 4
CHARACTERISTICS OF P-3 ORION ASW AIRCRAFT

Aircraft	Combat Radius (nautical miles)	Characteristics
P-3	2,070	A four-engined turboprop naval ASW aircraft, with sophisticated sensor and control equipment, magnetic anomaly detectors and a digital computer that integrates all ASW information. P-3s carry a variety of mines, depth bombs, torpedoes, nuclear depth bombs, sonobuoys, sound signals and massive markers. Some models are equipped with Bullpup missiles. Others have meteorological equipment for weather reconnaissance missions. Others have systems used for mapping the earth's magnetic field.

Source: *Jane's Fighting Ships 1974–75.*

determine patterns or operational relationships. When a definite pattern emerges from these data, combined with observations of Soviet naval operations in other waters and reports from other intelligence sources, this information goes into highly classified training manuals to help crews recognize specific threats and counter them.

Part of this task force is a squadron of four Asheville-class patrol boats that are armed with both radar-directed guns and the Standard surface-to-surface missile. The squadron is homeported at Naples with its own mothership, or tender. These ships displace only 245 tons, are 164 feet long and 24 feet wide, and have a crew of 27 men. They are powered by a combination of diesel and gas turbine engines, using the turbines to reach speeds in excess of 40 knots. This capability enables them to keep up with—or quickly get out of range of—any Soviet ship that may be sent to the Mediterranean.

Hunter-killer submarines. Task Force 69—the nuclear-powered hunter-killer submarine (SSN) force—has several duties. In case of war, its primary mission would be to protect the carriers and Polaris/Poseidon submarines. To do this the SSNs would try to eliminate any vessel—submarine or surface ship—that presented a threat. Once that was done, though it might be a matter of days or weeks if the conflict lasted that long, they would expand their target list to include enemy ships at greater distances from the carriers. In other words, the SSNs would join the rest of the fleet in establishing control of the sea.

The characteristics and operating tactics of SSNs, like those of SSBNs, are super secrets.[26] However, from reading *Jane's* and other published material one can get a glimpse of, if not a precise feel for, what these submarines are capable of doing. Take the Sturgeon class, for example. It is the largest class—37 vessels—of SSN. This type of submarine displaces 4,630 tons submerged, is 292 feet long, 31 feet wide and has a crew of 107. It has four torpedo tubes that can fire SUBROC (see Appendix Table A-3) and conventional and nuclear torpedoes. They are designed for sustained submerged operation at high speeds and substantial depths. *Jane's* lists 30 knots for a submerged speed but, like the SSBNs, they can probably go much faster.

---

[26] Even to go aboard one of these submarines in port is difficult, and to get permission to ride during a patrol is virtually impossible for a casual visitor. The problem, as a senior Sixth Fleet officer told me, is that during a patrol a SSN can be called on for an underwater tattletale mission at a moment's notice. The speeds and depths that it might reach are so highly classified—not to mention the nature of the mission itself—that a visitor would have to leave the submarine. By the time the visitor was put ashore or transferred to another ship the reason for the operation might have disappeared.

Their operating depth—a more significant capability—is harder to estimate. During World War II, German U-boats could dive below 600 feet without causing a fatal implosion (that is, the caving in of the structure under extreme water pressure).[27] France is building the Agosta class of submarine with conventional diesel/electric propulsion that is designed to operate at depths exceeding 1,800 feet.[28] The diving capability of the Sturgeon is probably in the area of 1,800 feet. The Los Angeles class of SSN, now just going into service, is designed for higher speeds and quieter operation, both of which characteristics help it escape detection. These vessels displace 6,900 tons, are 360 feet by 33 feet, and have a crew of 102.

This is the normal organization of the Sixth Fleet and the characteristics of the ships that are typically assigned to it. All of the task forces can be reinforced or reduced and others added depending on the political climate in the Mediterranean and elsewhere, on training requirements and, of course, on the nature of Soviet naval activity in the sea. There was a time when an antisubmarine warfare carrier (CVS) task force would operate in the Mediterranean and join the Sixth Fleet to conduct extensive training and intelligence operations. This task force, designated TF-66, was built around a converted attack aircraft carrier of World War II design and construction. These vessels carried about 2,400 men and 45 aircraft, including helicopters, all equipped with ASW equipment. However, ASW carriers are no longer used.

Political Uses of the Sixth Fleet

So far we have been concentrating on the technical characteristics of the Sixth Fleet; that is, what the fleet is designed to do on paper or what it can do in an environment where it is unopposed. But what about the fleet's political impact, its influence on the Mediterranean countries themselves? Any number of examples of the political use of the Sixth Fleet can be cited. Indeed, there is a political component to nearly all of its activities, for the fleet is used to fulfill or enhance the immediate and long-range international security objectives of

---

[27] For example, a German submarine, U-371, which was finally sunk on 4 May 1944 by a combined American, British, and French force off the Algerian coast, dived to 656 feet and came back to shallower depths to continue fighting. See Samuel Eliot Morison, *History of United States Naval Operations during World War II* (Boston: Little, Brown and Co., 1956), vol. 10, *The Atlantic Battle Won*, p. 256.

[28] The Agosta-class design capability was mentioned by Lawrence Griswold in "France: The Third Force?" *Sea Power*, vol. 16, no. 8 (August 1973).

the United States. It encounters opportunities to achieve these goals, but it also runs risks.

Port Visits. Take port calls, or "showing the flag" visits as an example. In addition to giving sailors some free time or "liberty" ashore, these visits can project a positive image and generate a certain amount of good will if they go smoothly. The very presence of a ship or group of naval vessels at anchor dressed with brightly colored signal flags or lights rigged fore and aft and smartly attired seamen is impressive. Often a variety of "people to people" programs are organized, including visits to the ships by local residents, band concerts in the town square, parties for orphans, blood donations to the local Red Cross, Red Crescent, or hospital, and so on. The hospitality is often returned by private groups and local officials who organize tours, folk festivals, or give receptions for the officers and men. The visits also bring spurts of spending in dollars that are welcomed by local businesses.

Sometimes, however, these port visits have been marred by drunken and disorderly behavior by U.S. sailors or by violent anti-American demonstrations. The conduct of sailors while on liberty has been the focus of increasingly critical attention. Bar fights and brawls over prices are a staple feature of the waterfronts of the world. Police and other officials who are responsible for enforcing the law in these areas have a fairly tolerant attitude towards these offenses, preferring to break up an argument or fight quickly, lock up the culprit until the next morning, and send him back to his ship. Problems arise when the violence involves major injury or death or large amounts of property damage, or when it takes place in neighborhoods away from docks and touches the ordinary citizen. When these incidents occur they are frequently reported in the press and used to criticize the presence of the Sixth Fleet and U.S. policy generally. The U.S. Navy recognizes the negative impact these incidents have had in the Mediterranean and has taken steps to reduce and control them.

Because port calls are so conspicuous they often become focal points for anti-American demonstrations or protests against the domestic policies of the local government. For example, violent riots in Istanbul and Izmir in 1969 and 1970, during which several American sailors were injured and thrown into the sea, caused a brief halt to Sixth Fleet visits to Turkish ports. When the visits cautiously resumed, the ships restricted their port calls to small, out-of-the-way coastal towns where they were least likely to encounter demonstra-

Table 5

SIXTH FLEET VISITS IN THE MEDITERRANEAN,
1964 AND 1974

1964	1974
France	Cyprus
Gibraltar	Egypt
Greece	France
Italy	Greece
Lebanon	Italy
Libya	Morocco
Malta	Spain
Morocco	Tunisia
Spain	Turkey
Tunisia	Yugoslavia
Turkey	

Source: U.S. Navy Office of Information.

tions. Martial law was imposed in Turkey in March 1971, and by March 1972 the domestic situation had calmed enough for larger port cities to be put back on the Sixth Fleet's list.[29]

Throughout the Mediterranean the possibility of terrorism adds another, more dangerous dimension to the concern for politically motivated violence against Sixth Fleet ships and men. One of the constant worries is that an extremist guerrilla group may attempt to launch a rocket or otherwise sabotage a U.S. Navy vessel. Stringent procedures are followed when this is considered a real threat and generally tighter security has been enforced with the spread of political terrorism.

The Sixth Fleet's access to ports in the eastern and southern Mediterranean has diminished in the last decade. This is due in large part to Arab hostility toward the fleet in particular and United States policy in the Middle East in general. Before the 1967 Arab-Israeli war, for example, Sixth Fleet ships of all types made regular calls at Beirut. These visits were gala affairs, occasions for formal receptions and parties for officers and sports contests between ships' teams and local athletic clubs. Since that war, Lebanese ports have been closed to the Sixth Fleet. In 1963, American warships called at both Israeli and Libyan ports. Today, such visits are politically impossible.

[29] *New York Times*, 15 March 1972.

Countries in other parts of the sea fear the possibility of an Arab oil embargo or simply of tougher terms on buying petroleum should they appear to be too closely identified with the Sixth Fleet. This was one of the concerns of Greece during the homeporting negotiations and a factor that contributed to the end of the arrangement.

Domestic political factors and the international climate are major considerations bearing on the timing and tone of port calls. The process of drawing up the quarterly schedules for visits of some 45 ships is a juggling operation designed to please everyone without offending anyone. This task is the full-time occupation of a Sixth Fleet staff officer and involves the American embassies in each of the countries to be visited, the respective foreign and defense ministries, local police and security forces, and in the United States the Pentagon and Department of State.

Bearing in mind the need to keep a balanced deployment of ships in both basins and the fact that some countries will not permit visits by American warships at all, the scheduling officer asks the American naval attaches in the remaining countries for their suggestions on the number, frequency, timing, and type of ships for port calls for the next quarter. Each naval attache normally jots down his suggestions and passes them on to the embassy's political section. Officials in the country's foreign and defense ministries are asked about the advisability of visits and whether or not they have any preferences about dates, choice of port, and number of ships. This information is folded into judgments by embassy officials. If the domestic political scene is generally quiet and is judged likely to remain so during the next quarter, the suggested dates are reviewed closely to see if the proposed visit coincides with an election, national holiday, or local saint's day in the port town where the ship would anchor. Depending on the character of the event a visit by a Sixth Fleet ship could be inappropriate or tax the resources of local law enforcement authorities. The embassy could also ask that a ship with relatively few men be sent or that an auxiliary ship like a supply vessel rather than, say, a destroyer make the visit.

The proposed itinerary is sent to the country desk officer in the State Department and to the Pentagon for additional suggestions and comments. For example, a high-ranking foreign official might have suggested during a visit to Washington or some other capital that two or three Sixth Fleet ships would be welcome in his country at the end of a joint naval exercise. This request could result in such a visit or, if it is learned that a debate on a major foreign policy issue is likely to be held during the same period, the proposed visit might be deemed inadvisable. The State Department, particularly its younger officers, tends to be conservative in giving clearances for port visits, while the U.S. Navy is more assertive and would like to make wider use of the visits. Once the schedules are approved they are adhered to with only minor variation, though an international or domestic crisis could alter them at any time.

Rescue Operations. Occasionally the Sixth Fleet is asked to take part in rescue efforts at sea or render aid during a natural disaster. In March and April 1973, for example, following devastating floods in Tunisia, men and aircraft from the aircraft carrier *Forrestal*, the amphibious ship *Ponce*, and the guided missile destroyer *Sampson* assisted in rescue operations and dropped supplies to isolated areas. More than 1,000 stranded Tunisians were plucked by helicopters from roofs, treetops, and high ground. Many thousands of others received food, water, blankets, and medical supplies from the Sixth Fleet. During a widely publicized ceremony following the operation, Tunisian President Habib Bourguiba called the Sixth Fleet "a friend we can count on in times of need." [30]

Suez Canal Clearing Operation. A rather unique application of American naval power occurred during 1974 when the U.S. Navy participated in the effort to clear the Suez Canal. Beginning in April, a Mine Counter-Measures Task Force, designated TF-65 for the duration of the operation, joined the Egyptian, British, and Soviet navies in removing mines and other explosives.

The operation itself, which was staged from the helicopter carrier *Iwo Jima* at anchor off Port Said, used RH-53D Sea Stallion helicopters that are designed for aerial mine sweeping. These helicopters detect, locate, and detonate sunken mines. The mines they were searching for were mostly American-made weapons that had been supplied to Israel. Israel provided information on their approximate location and the U.S. Navy pinpointed them for the Egyptian frogmen to remove. More than 800 U.S. servicemen worked on the project.

In the context of warming American-Egyptian relations and the need for improving the image of the Sixth Fleet in the Arab world, the mine-clearing operation had several benefits. First, the task force performed a service of obvious benefit to Egypt, to the Mediterranean, and to world shipping. Second, it provided a rare opportunity for contact on a working level between the American and Egyptian navies and helped to remove the stigma attached to the Sixth Fleet. Third,

---

[30] Ibid., 8 April 1973.

the operation resulted in further fleet visits, including a call by the flagship (29 July–2 August), the first visit to Egypt by a Sixth Fleet flagship since 1962. During former President Nixon's visit to Egypt (12–14 June 1974), the *Iwo Jima* sailed to Alexandria and provided helicopter transportation for the President between his guest house and official appointments. A navy band provided some of the music for social functions during the state visit.

Lebanon Landings. A more dramatic use of the fleet as an instrument of diplomacy was the landing of marines in Lebanon and the naval build-up in the eastern Mediterranean in July 1958. That operation was part of a much wider response to what appeared in Washington to be a bold move by the Soviet Union to penetrate the Middle East and seriously endanger vital American interests. A comprehensive study of that era would fill a separate book. What concerns us here is how the United States used the Sixth Fleet and other military power in the Mediterranean to react to that threat.

In mid-1958 the United States and, to a lesser degree, Britain were deeply involved in a delicate political balancing act. On the one hand they were subtly encouraging Arab nationalism as a force to check the spread of communism and Soviet influence. American policy makers reasoned that it was better that Syria join with Egypt in the United Arab Republic than that Damascus be the seat of a pro-Soviet government. At the same time, the United States and Britain were trying to contain the new nationalism within certain limits to prevent it from toppling the pro-Western regimes in Lebanon, Jordan, and Iraq. By June, a full-scale civil war was raging in Lebanon and there was dangerous tension in Jordan. Both were being encouraged by broadcasts from Cairo and Damascus. Iraq, the two countries felt, was firmly in the hands of a government formally allied with the West.[31] Just off stage, the Soviet Union was giving active sympathetic propaganda support to any anti-American or anti-British efforts in the Arab world. When the unexpected news was flashed to Washington and London on 14 July 1958 that an anti-Western faction of the Iraqi army had seized power in Baghdad, assassinating government leaders and the heads of the royal family, Lebanon and Jordan appeared to be in the path of a "new Middle East hurricane."[32]

---

[31] The Baghdad Pact was concluded in 1955 with Turkey, Iraq, Iran, Pakistan, and Britain. Though the United States was a de facto member, it did not formally join the alliance, hoping to retain a measure of influence with Nasser who bitterly denounced Iraq's participation.

[32] J. M. MacKintosh, *Strategy and Tactics of Soviet Foreign Policy* (New York: Oxford University Press, 1967), p. 234.

The United States told government leaders in Beirut that an urgent appeal for troops would be answered on the basis of the Eisenhower Doctrine's rationale for repelling outside aggression. The day of the Iraqi coup, Lebanese President Chamoun issued such a call for help. By three p.m. the next day, U.S. Marines were landing on the beaches around Beirut. Within the next 72 hours a total of 5,400 marines were put ashore, an American airborne battalion had secured the city's airport and a 50-ship U.S. naval force, including two aircraft carriers, was concentrated off Lebanon's coast. While American forces were landing in Lebanon, the British had flown 2,000 paratroopers to Amman in response to a similar request for assistance from King Hussein. Almost immediately the political violence ceased in both countries and the situation in Lebanon and Jordan was stabilized.

Iraq was a different matter. The country appeared to be lost to the Western camp but it was not clear in Washington or London at the time whether the new regime was Iraqi nationalist, pro-Nasser, or pro-Soviet. There was nothing to suggest that the Soviet Union had had a hand in engineering the coup. Moscow did not have a diplomatic mission in Baghdad and the Communist party was outlawed by the monarchy. But the behavior of the army officers was enough to alarm the West. Diplomatic relations were restored with the Soviet Union and friendly overtures were made to Egypt in the first days after the coup. The United States and Britain wanted to see to it that friendship between Iraq and these two countries went no further.

To prevent any further tilt of Iraq toward Moscow or Cairo would require a stronger response than the show of force in the eastern Mediterranean. The marine landings and the naval reinforcements were part of a worldwide alert of American forces. Though unannounced, the alert was clearly visible. Communities around Strategic Air Command (SAC) bases in the United States knew that the men who flew the giant B-52s had reported to duty in the middle of the night and did not return home for several days. Auto rental firms depleted their fleets of cars so that B-52 crews could quickly shuttle between the dining halls and the planes where they slept.

Meanwhile, on 18 July, the Soviet Defense Ministry announced that military maneuvers involving air and airborne units were to be staged in the Trans-Caucasus and Turkistan military districts and in Bulgaria—in other words, on the line across the northern frontier of Turkey and Iran across which Russian troops would have to move if they were to head for Iraq.[33] Mob violence continued for several days in Baghdad and the leadership of the new government remained uncertain until 27 July, when Abdul Karim Kassim announced that Iraq was to be "part of the Arab nation" but an "independent sovereign Republic." This was a clear sign that Iraq would not be joining the United Arab Republic or the Soviet bloc. That situation, too, stabilized. There was a net loss to the West, but a loss that could have been worse had it not been for a dramatic show of force that included naval power.

1970 Jordan Crisis. The crisis during the September 1970 civil war in Jordan provides another example of the use of American naval power to influence events on land. In this instance, however, there was no direct intervention. Although the country involved has no Mediterranean coastline, it is part of the larger geopolitical Mediterranean area. In many ways the regional Middle East situation in 1970 was similar to that of 1958. There was more at stake for the United States than just the survival of King Hussein's staunchly pro-American regime, and the concentration of naval forces in the eastern Mediterranean was only part of a wider military and diplomatic reaction to end the violence in Jordan and reduce the threat of violence throughout this region. It is necessary to look at the background and the crisis in some detail in order to understand the role of naval power in our broader context.

The year began with a general fear of a new Arab-Israeli war. Israeli jets had virtual control of Egyptian airspace, with some Israeli planes striking within five miles of Cairo. These "deep penetration" raids stopped after the unprecedented rush of Soviet air defense equipment to Egypt. The build-up included modern MiG-21J interceptors, an interlocking radar-controlled antiaircraft gun and missile system, and Soviet pilots and technicians to fly the planes and operate the ground defenses. It was the first time such a large number of Soviet military personnel and such sophisticated Soviet hardware had been deployed outside the Warsaw Pact area.[34] As the danger of a new Arab-Israeli war increased, the likelihood of a direct superpower confrontation grew as well. Recognizing this, both the United States and the Soviet Union convinced their clients to pull back from the brink. After protracted negotiations which put intense pressure

---

[33] Ibid.

[34] By 31 December 1970 when the build-up peaked, there were between 12,000 and 15,000 missile crewmen, more than 200 pilots, and 150 MiG-21J interceptors in Egypt. In addition, the Soviets controlled six airfields. *Strategic Survey 1970* (London: International Institute for Strategic Studies, 1971), p. 47.

on both Israel and Egypt, the two sides agreed to a cease-fire and stand-still accord on 7 August.

Meanwhile, the Palestinian guerrillas, by this time openly challenging government authority in Jordan, were becoming increasingly militant and calling for more military pressure on Israel. The cease-fire agreement along the Suez Canal literally pulled the rug out from under them, leaving many of the guerrillas bitter and determined to try to destroy the truce. The response of the Popular Front for the Liberation of Palestine, one of the more extremist groups, was to hijack three foreign airliners to Jordan on 6 and 9 September and to hold the passengers hostage despite government attempts to secure their release. The events were an open affront to King Hussein's personal prestige. They were also the latest in a series of escalating challenges to his authority. The army repeatedly had urged the king to take decisive steps to suppress the guerrillas. The United States, too, tried to encourage Hussein to take firm measures against the guerrillas whose activities it viewed as an obstacle to American efforts for a settlement. The spectacle on 12 September of the planes exploding in the desert convinced the king to act.

Early in the morning on 17 September, Jordanian tanks rolled into Amman and troops began systematically attacking guerrilla strongholds in the city. Resistance, however, was much heavier than expected because the guerrillas were well armed with automatic rifles, machine guns, and anti-tank weapons. Another factor in the slow pace of the assault was that the guerrillas were fighting on familiar ground while the government forces were trained for armored tactics in the desert.

Even before the actual fighting started, Washington had several key concerns. First, 38 Americans were among the 54 hostages still held by the guerrillas somewhere in Amman. There were 450 other Americans resident in Jordan. Second, the responses of Syria and Iraq were worrisome. Syria, which had refused to join in the latest round of peace diplomacy, denounced the cease-fire and generally supported the guerrilla position. Syria had also provided assistance to the guerrillas during earlier clashes with Hussein's troops. Iraq had about 10,000 troops stationed in Jordan, although at some distance from Amman. Their intervention could pose a threat to Hussein. Third, there was a general feeling that if Hussein did not move against the guerrillas he would probably not have a kingdom very much longer. One factor in this assessment was the belief that Israel would not tolerate a guerrilla takeover in Jordan and would take military action, for example, by crossing over the Jordan River cease-fire line.

If Israel intervened, the fragile cease-fire along Suez would crumble. And when the inevitable civil war erupted in Jordan, there was the danger of a wider conflict that would also destroy the cease-fire. The problem for Washington was how to save Jordan and the Egyptian-Israeli cease-fire as well.

Because of the seven-hour time difference between Jordan and the east coast of the United States, the news of the fighting was widely broadcast on morning radio and television programs on 17 September. That same day, after President Nixon held an off-the-record press conference in Chicago, one afternoon paper wrote: "The United States is prepared to intervene directly in the Jordan civil war should Syria and Iraq enter the conflict and tip the military balance against the government forces loyal to King Hussein." [35] There was some confusion, however, about American intentions. The paper did not quote the President directly, though the words were clearly Mr. Nixon's. The White House asked the paper to remove that portion of the story from later editions, and White House statements in response to reporters' questions the next day were vague. A much clearer statement of American intentions came on 17 September from Secretary of Defense Melvin Laird, who said that the United States was "prepared to evacuate Americans from Jordan if necessary" and that the helicopter carrier *Guam* with a U.S. Marine contingent was steaming for the eastern Mediterranean. But after the White House had backed off from the *Chicago Sun-Times* story, it was not clear whether Laird's remarks still held.

While the Sixth Fleet was being sent to the eastern Mediterranean, other American units were put on alert for possible use in Jordan. These included the 15,000-man 82nd Airborne Division at Fort Bragg, North Carolina, and the 3rd Regiment of the 1st Infantry, consisting of some 3,500 men based in Germany.

On 19 September, tanks from Syria bearing the markings of the Palestinian Liberation Army, an organic part of the Syrian armed forces, began spearheading a guerrilla push to the south toward Amman to relieve other irregulars fighting in the city. By 21 September, the guerrillas' radio, broadcasting from Damascus, claimed that their forces controlled a vital road junction north of Amman and were within 50 miles of the capital.

On 20 September, Secretary of State William Rogers dispelled any doubts about the U.S. position. "We condemn this irresponsible and imprudent intervention from Syria into Jordan. This action carries with it the danger of a broadened conflict." The warning was

---

[35] *Chicago Sun-Times*, 17 September 1970.

given to the Soviet chargé d'affaires in Washington by Assistant Secretary of State Joseph Sisco who told the diplomat that the United States hoped the U.S.S.R. would use its influence with Syria to end the incursion. The next day, the State Department announced that it had been informed by the Soviet Union that contact had been made with Syria, but added that "the real test . . . will turn on whether the forces from Syria are withdrawn from Jordan." By 22 September they had not been withdrawn.

The role of the Soviet Union deserves mention. From the beginning of the crisis, Moscow, while making noises in support of the guerrillas, had made it clear that it would not intervene and had urged that other powers stay out as well. The warnings appeared to be directed more at Syria and Iraq than at the United States. The U.S.S.R. might have wanted a neutral or pro-Soviet government in Jordan, but it was not willing to confront the United States to get it or upset the cease-fire along the Suez Canal it had recently worked to achieve.

In Jordan the situation was desperate. Spearheaded by the tanks, the guerrillas now controlled Irbid, the country's second largest city, and were pushing even closer to Amman. Hussein's army was tied down in the city, leaving the north to the guerrillas and to the tanks from Syria. The Iraqi forces did not get involved. Hard pressed, Hussein put out feelers to the big powers, though it was never clear in public exactly what form of assistance he was seeking. Privately, the king asked the United States if he could count on help from the Sixth Fleet in the form of air cover so that his armor could move to the north. So far, the small Jordanian air force had not been used in the north because of the certainty that the 11 planes in its inventory would be overwhelmed by Syria's 145 combat aircraft. But without air cover Jordanian tanks would be sitting ducks. Hussein needed some assurances from the United States, and such assurances were given through private channels.

At sea, by 21 September Sixth Fleet pilots had already been briefed on possible targets in Syria, the nature of their defenses, and the area of the fighting in Jordan. Plans were completed for the evacuation of American citizens using marines flown in by helicopter to secure Amman airport and to bring in army units if needed. Israel was prepared to grant overflight rights for the aircraft. The ability to conduct such an operation was clearly apparent to the Soviet Union whose fleet was monitoring every move by the Sixth Fleet. No doubt this information was relayed to Damascus.

The main American naval force was cruising north of Egypt ju southwest of Cyprus, about 250 miles west of Israel. It consisted two attack aircraft carriers, the *John F. Kennedy* and the *Saratog* the helicopter carrier *Guam*, one cruiser, the flagship of the flee and ten destroyers, some of which were armed with SAMs. Oth American ships were in the area. The carrier *Independence*, its d stroyer escorts, and an amphibious task force were near Athens. The capability to intervene was on the spot, with adequate reinforc ments close by.

By 23 September, the Jordanian air force was in action, strafin the tanks now retreating into Syria. The Syrian air force did n approach the area of the fighting. By 25 September all of the tank from Syria had been either destroyed or withdrawn and mopping u operations around Irbid and Amman were nearly completed. All b 6 of the 38 American hostages were released on 26 September an flown to Nicosia, Cyprus, the following day. The crisis had ease

Having studied these examples of the political use of America naval power, let us inquire, how does the Sixth Fleet react to its ma adversary in the sea, the Soviet Mediterranean fleet? How does th Soviet Union react to the Sixth Fleet? And how are the superpowe likely to react to each other in a conflict? Before answering thes questions, we must examine the Soviet military presence in th Mediterranean.

---

[36] Admiral Isaac C. Kidd, Jr., "View from the Bridge of the Sixth Fleet Flagship, *United States Naval Institute Proceedings*, vol. 98, no. 2/828, p. 28.

4
THE SOVIET PRESENCE

Overview and Background

The Soviet Mediterranean fleet, the Fifth Escadra,[1] which has been patrolling that sea in strength since 1967, is a powerful, modern naval force with some weapons that are not possessed by the Sixth Fleet. But compared to the U.S. presence in both tactical and strategic terms, the Soviet presence, which is almost all naval, is less extensive and less capable, primarily because at present the Soviet Union does not have air power based in the Mediterranean. Russia is expected to overcome this deficiency soon, however, and thus to dramatically alter the psychological balance of power. Its first fixed-wing aircraft carrier is near completion in the Black Sea and another is under construction there. Even though these new ships eventually may be deployed elsewhere, their first cruises in foreign waters must necessarily be in the Mediterranean.

The key to understanding why there is a Soviet presence in the Mediterranean and why Russian policy is aimed at expanding that presence is largely a matter of geography and history. The Mediterranean is the gateway to the Black Sea and to one of the major industrial and agricultural regions of the Soviet Union: the Ukraine, which abuts on the Black Sea, is not only the chief wheat-producing area, Russia's breadbasket, it also accounts for 50 percent of the pig iron produced in the Soviet Union, 40 percent of the steel, 35 percent of the manganese, and 34 percent of the coal. The ports and large population centers of Odessa, Nikolayev (also a major ship-building complex), Sevastopol, Yalta, Kiev, Rostow, and Volgagrad (formerly Stalingrad) either border the Black Sea or are directly accessible

---

[1] In the Soviet navy an escadra is one notch below a fleet in rank.

from the inland transportation arteries of the Dnieper and Don rivers that flow into the Black Sea. Also, there is continuing concern in Moscow about the political reliability of the Ukraine, as well as of the neighboring Balkans, particularly Rumania. Together these regions could be called the soft underbelly of the Soviet Union.

The other dimension of anxiety about this rich, strategic region is that throughout history the Turkish Straits and Black Sea have been avenues for military thrusts against the Russian heartland. This was true in the Crimean Wars, for example. However, the Soviet Union does not control the area's only access route, the Turkish Straits. Indeed, much to its annoyance, the country which does control the straits, Turkey, is a member of NATO, an anti-Soviet alliance. Further, since the end of World War II, the Mediterranean has been one of the main staging areas of Western military power targeted against the Soviet Union. All of this helps to explain why Russia—both tsarist and Soviet—has consistently made breaking out of the Black Sea to the warm waters of the Mediterranean one of the objectives of its foreign policy.[2] Its most successful means of pursuing this end has been a permanent naval presence in that sea that can be used as a tool of diplomacy.

Let us look at one important Soviet statement on the subject of the Mediterranean and naval power to get a clearer understanding of the view from Moscow. Writing in the Soviet Defense Ministry newspaper *Red Star*, then Vice Admiral N. I. Smirnov said:

> . . . the Sixth Fleet is stationed in this area, thousands of miles away from U.S. shores, in order to control lines of communications passing through the zone of the Black Sea Straits and in the Mediterranean Sea, as well as to be prepared to strike at enemy targets from sea and air with

---

[2] Active Russian interest in the Mediterranean began in the mid-eighteenth century when Moscow sought ways to outflank Turkey. As early as 1770, Russian warships operated in the sea and, at various times until 1807, Russian troops occupied Corfu and the Ionian Islands as well as several Dalmatian ports and dominated the eastern Mediterranean with garrisons at Cyprus, Beirut, Latakia, Jaffa, Gaza, and Alexandria. When its fleet was unable to force the Turkish Straits in 1807, the Russian navy had to leave the Mediterranean through Gibraltar and was captured by the British near Lisbon. There is more than just historical irony or coincidence in the comparison of tsarist and Soviet interests in the Mediterranean because many of the same places are the focus of Moscow's current foreign policy. For further references see: Lawrence L. Whetten, *The Soviet Presence in the Mediterranean* (New York: National Strategy Information Center, Inc., 1971); Norman E. Saul, *Russia and the Mediterranean 1797-1807* (Chicago: University of Chicago Press, 1970); and Vali, *The Turkish Straits and NATO*.

nuclear and conventional weapons, and the Soviet Union and other socialist countries are designated as the principal enemy. . . .

In these circumstances, the interests of security have demanded that the Soviet Union indefatigably strengthen its defense capability. Our state, which is, as is known, a Black Sea and consequently also a Mediterranean power, could not remain indifferent to the intrigues of those fond of military ventures organized directly adjacent to the borders of the U.S.S.R. and other socialist countries. No one can be allowed to turn the Mediterranean into a breeding ground of a war that could plunge mankind into the abyss of a worldwide nuclear-missile catastrophe. The presence of Soviet vessels in the Mediterranean serves this lofty, noble aim.[3]

When and how did this naval build-up begin? What form does this presence take today? What types of weapons are involved and what are their capabilities? Where are Soviet forces located? How does the Soviet Union use its navy as a political instrument? And how could Soviet naval presence develop in the future?

The first major deployment of Soviet naval forces to the Mediterranean after World War II occurred in 1958 shortly after large numbers of A-3Ds, a long-range jet aircraft capable of carrying nuclear weapons, had been assigned to Sixth Fleet attack carriers. To counter this threat, the Soviet Union moved eight Whiskey-class submarines and a submarine tender to a new naval base at Vlone, Albania. About the same time, Soviet intelligence-gathering ships (AGIs) began to patrol the sea regularly.[4] This reaction reflected the great concern of the U.S.S.R. In 1960 these submarines along with AGIs conducted an extensive exercise in the Aegean Sea that appeared to be based

---

[3] *Red Star*, 12 November 1968, translated in *Current Digest of the Soviet Press*, vol. 20, no. 47 (11 December 1968), pp. 7-8. This view takes on added significance because the author, who has served in the Mediterranean and commanded the Soviet Pacific Fleet, is now first deputy commander in chief of the Soviet navy, a position equivalent to the U.S. vice chief of naval operations. He is now in line to be the next head of the Soviet navy.

Just the titles of two articles in the *Soviet Military Review*, an English-language periodical printed in Moscow, provide additional insights into Soviet thinking: "The Mediterranean Is Not an American Lake," by then first deputy commander in chief of the Soviet navy, Admiral of the Fleet V. Kasatonov, January 1969, and "NATO Threat in the Mediterranean," Colonels G. Arzumanov and V. Katerinich, May 1971.

[4] Nicholas George Shadrin, "The Development of Soviet Maritime Power" (Ph.D. dissertation, George Washington University, 1972), vol. 1, p. 226.

on scenarios of attacking U.S. aircraft carriers moving into the Aegean to launch a nuclear strike against the Soviet Union.[5]

This sort of tit-for-tat reaction, which would become a pattern, indicated Moscow's great apprehension over American naval activities in the Mediterranean. It probably had a long-range impact on the Soviet Union's shipbuilding program and naval strategy. The Soviet leaders chose to try to nip the threat in the bud, so to speak, by stopping the attack before it got under way.

The unceremonious break between the U.S.S.R. and Albania in May 1961 and the abrupt loss of a submarine base and two submarines were serious setbacks for the Soviet Union in the Mediterranean. Diesel/electric submarines require extensive support facilities to operate for long periods away from home waters. At the time the Soviet navy did not have sufficient numbers of long-range surface ships or submarines to divert to the Mediterranean. Submarines have to make the long journey from the Baltic or Northern seas because they are prohibited by the Montreux Convention from entering the Mediterranean for combat patrols from the Black Sea. (Submarines from the Black Sea have occasionally operated in the Mediterranean but they have changed their hull numbers and declared shipyards as their destinations to avoid directly challenging Turkey on the terms of Montreux.)

So between 1961 and early 1964, Soviet naval activity in the Mediterranean was spotty. Another reason for the reduced presence during this period could be that Moscow considered the defense of the northern and Baltic maritime borders of the U.S.S.R. as more vital.[6] (This period was also one of high tension in Central Europe with the construction of the Berlin Wall in August 1961, and included the Cuban missile crisis in 1962.)

The date that most authorities agree on as the beginning of a permanent Soviet naval presence in the Mediterranean is June 1964, though present force levels were not reached until 1967. During 1964, 15 Soviet warships patrolled the sea and the commander of the Black Sea fleet, Admiral S. Y. Chursin, made an official visit to Yugoslavia.[7] All of this followed the introduction of American Polaris submarines into the Mediterranean in 1963. This event probably more than any other had a profound influence on the subsequent

[5] John Broderick Chomeau, "Seapower as a Political Instrument: The Soviet Navy in the Mediterranean" (Ph.D. dissertation, University of Notre Dame, 1974), p. 72.

[6] Ibid., p. 74.

[7] Ibid., p. 75.

shape of the Soviet military presence in the Mediterranean. It caused the initial heavy emphasis on antisubmarine forces in the Mediterranean and led to a briefly successful search for naval and air bases in Egypt, a search that will continue.

One analyst, George S. Dragnich, has said, for example, that the introduction to the Mediterranean of the Polaris missile submarine

> seriously undermined the Soviet Navy's attempt to establish meaningful forward deployments, since its existing resources were still insufficient to check even the lesser strategic threat posed by the Sixth Fleet's attack aircraft carriers. In order to maintain a permanent naval presence in the Mediterranean to meet this challenge (both SSBNs and CVAs) the USSR needed access to naval facilities in the region itself—at least until its hard-pressed naval expansion program could render such facilities redundant.[8]

The Fifth Escadra

There is very little in the public record about the Fifth Escadra's organization. It is reasonable to assume that it is divided into subcommands by function—surface combat, submarine, amphibious, and supply and replenishment. The commander, usually a rear admiral, normally keeps his flag and staff aboard a submarine tender that frequently visits Tartus, Syria. It is known that the commanders of the surface combat and submarine forces also have their headquarters aboard the same vessel. Most analysts feel that a submarine tender is used as a flagship and base for the staffs of other commanders because of the living and office space that such large ships can provide. The escadra commander, however, often cruises aboard other ships with a portion of his staff to observe particular training exercises.

The size of the escadra can, of course, vary. There are normally about 55 Soviet naval vessels in the Mediterranean, of which 20 to 25 are warships. A typical distribution would be: 2 cruisers, 4 destroyers, some armed with surface-to-air missiles, 12 submarines, including nuclear-powered vessels, 4 amphibious ships with naval infantry (Russian marines), and an assortment of tenders, oilers, supply ships, and intelligence-gathering vessels.

---

[8] See George S. Dragnich, "The Soviet Union's Quest for Access to Naval Facilities in Egypt Prior to the June War of 1967," *Soviet Naval Policy: Objectives and Constraints*, ed. Michael McCGwire, Ken Booth, and John McDonnell (New York: Praeger, 1975), p. 252.

Like the U.S. Sixth Fleet, the Soviet Mediterranean squadron can include any combination of ships. It might be useful, though, to give a thumbnail sketch of the characteristics of the major classes of Soviet ships that have been regularly seen in the Mediterranean in order to assess the threat faced by the Sixth Fleet. Such a sketch will demonstrate that Soviet designers tend to pack more weapons systems into a single hull than do American naval engineers and that many Soviet ships carry weapons that do not exist in the U.S. Navy.

Helicopter Carriers. Often the Soviet Mediterranean squadron includes a Moskva-class antisubmarine helicopter carrier, one of the most impressive warships of its kind in the world. It looks like a cross between a cruiser and aircraft carrier. Though not as versatile as an aircraft carrier because it does not carry fixed-winged planes, the Moskva can perform a wide variety of missions and in this its design is typical of many Soviet naval vessels. In a hull of 625 by 115 feet, displacing 18,000 tons fully loaded, are contained an imposing array of weapons including 18 ASW helicopters, hull-mounted and variable-depth sonars, antisubmarine rockets, torpedoes, SAMs, and a complement of electronic and communication equipment sophisticated enough to enable the Moskva to serve as a command ship. It has a crew of 800 and its geared turbine engines give it a speed in excess of 30 knots.

The helicopters, the Kamov KA-25 "Hormone," are stored in a hangar located under the 295-foot flight deck at the stern of the ship. These aircraft have a cruising range of 351 nautical miles at a speed of 104 knots. They are fitted with radar and dipping sonar and have both internal and external storage for torpedoes, depth charges, and flares and other markers. There are two such ships in the Soviet navy, the *Moskva* and the *Leningrad*. Though they are called ASW cruisers in the Soviet list of ships, the Western designation is helicopter missile cruiser, which better sums up their capabilities.[9]

Cruisers. The Kara class of cruiser, which was first seen in public when it entered the Mediterranean from the Black Sea on 2 March

---

[9] *Jane's Fighting Ships 1974-75*, for example, says of the Moskva class: "Alongside what is apparently a primary A/S role these ships have a capability for A/A (anti-aircraft) warning and self-defense as well as a command function. With a full fit of radar and ECM equipment they clearly represent good value for money" (p. 533).

Another authority on the Soviet navy, Nicholas George Shadrin, writes in "The Development of Soviet Maritime Power," p. 124: "In certain areas, particularly in such confined basins as [the] Mediterranean (where the ship has been employed) Moskva might have a certain marginal anti-Polaris capability."

1973, is another impressive example of Soviet warship design. It carries nearly every conceivable type of naval weapon except fixed-wing aircraft. The *Nikolayev*, the first ship of this type, is fitted with 8 surface-to-surface missile (SS-N-10) launchers, 8 surface-to-air missile (4 SA-N-3 and 4 SA-N-4) launchers, 8 radar-directed guns, 48 multi-barrel units for antisubmarine rockets, 10 torpedo tubes, hull-mounted and variable depth sonars, and 1 antisubmarine helicopter. It displaces 9,500 tons, measures 570 by 60 feet, and is powered by gas turbines giving it a speed of about 34 knots. "The combination of such a number of systems presents a formidable capability, matched by no other ship," says *Jane's Fighting Ships*. This ship is "clearly capable of prolonged operations overseas." [10]

Two other types of Soviet cruisers often seen in the Mediterranean are the Kresta II and Kynda classes. The Kresta II, like the Kara class, carries a wide array of weapons, though it is slightly smaller. The Kresta II displaces 7,500 tons and has eight SS-N-10 and four SA-N-3 launchers, plus ASW weapons and radar-directed guns. It can carry two Hormone helicopters. The Kynda is even smaller and has a slightly less complete weapons package. Its eight surface-to-surface missile launchers fire the longer range SS-N-3 and have a limited reload capability.

An older type of cruiser is the 12-ship Sverdlov class built in the early 1950s. It displaces 19,200 tons, measures 689 by 72 feet, and is heavily armored. It has 12 six-inch long-range guns. The conventionally designed Sverdlov is no longer considered a modern ship of the line in the Soviet navy, though some have been converted to carry surface-to-air missiles and many remain in service as alternate command vessels. In this respect the Sverdlov class is similar in design and concept to the guided missile cruisers used as Sixth Fleet flagships, though the Soviet Union does not appear to use them as command ships in the Mediterranean. The Sverdlov also as a mine-laying capability.

Destroyers. In the category of smaller ships—destroyers, escorts, and corvettes—several classes must be mentioned. The first is the Krivak-class destroyer displacing 5,200 tons and driven by gas turbine engines giving it rapid acceleration and high speed. It is very similar in design to the larger Kara-class cruiser in that it contains a strong and varied complement of weapons, including surface-to-surface and

[10] Moore, *Jane's Fighting Ships 1974-75*, p. 550.

surface-to-air missile launchers, a VDS with associated MBUs and two banks of torpedo tubes.[11]

Another class of destroyer that is frequently seen in the Mediterranean is the Kashin class. This type of ship, which displaces 5,200 tons, is powered by gas turbine engines and has four SAM launchers plus an ASW capability and conventional guns. The Kotlin-class destroyer, also a familiar sight in the Mediterranean, displaces 3,885 tons. Many are armed with one twin SAM launcher. All have a suit of guns and antisubmarine weapons. A much smaller ship is the Nanuchka class of missile corvette. Displacing only 800 tons, this ship has six SS-S-9 launching tubes and a twin SAM launcher. The SS-N-9 surface-to-surface missile has a 150-mile range at supersonic speeds.

Submarines. Of great concern to Western naval officers in the Mediterranean are the large numbers of Soviet submarines. The Soviet submarine force is the largest in the world, and the submarines that patrol the Mediterranean present a formidable threat to the Sixth Fleet and Western navies both in number and sophistication. They are basically of two types: the conventional or nuclear-powered cruise missile submarines (SSGN and SSG) which are armed with torpedoes and whose primary role is to sink or otherwise incapacitate aircraft carriers and surface warships and conventional and nuclear-powered antisubmarine submarines that are primarily designed to destroy American Polaris/Poseidon submarines.

The most significant example of the first type is the Charlie class (a NATO designation) of submarine that is nuclear powered and has the capability of launching a short-range supersonic antishipping missile while submerged. *Jane's Fighting Ships* states that the Charlie class "must be assumed to have an organic control for their missile system and therefore pose a notable threat to any surface force." [12] The submarine is 295 feet long by 33 feet and displaces 5,100 tons when submerged. Because it is nuclear powered its speed is thought to be in excess of 30 knots and its combat endurance, like that of other nuclear-powered ships, is virtually unlimited for all practical purposes. Its crew is thought to be about 100 men.

Its weapons system consists of eight torpedo tubes and eight SS-N-7 missile tubes. And it carries nuclear as well as conventional torpedoes. The same nuclear capability probably exists for the missiles which are fired from launching tubes near the bow. The

---

[11] Ibid., p. 554.
[12] Ibid., p. 539.

ability to fire these tactical missiles submerged is the unique quality of this submarine, though because the SS-N-7 has a range of only 26 nautical miles, it risks being detected by a carrier's ASW defenses. Balancing that, however, are the quiet hull and engine design as well as the speed of this submarine, which may enable it to escape destruction until after its missiles are fired.

Another type of submarine that has been seen frequently in the Mediterranean is the Echo II class. This vessel is also nuclear powered and can fire its missiles, which are capable of containing nuclear warheads, in excess of 400 miles from their targets. The Echo II displaces 5,600 tons, is 387 by 28 feet, and has a crew of about 100. It has ten torpedo tubes, four of which are at the stern, and eight launching tubes for the SS-N-3 antishipping missile. *Jane's* lists its speed as only 20 knots. While the great range and nuclear capability of its missiles are substantial assets, the ship must surface to fire them.

The Juliet class of submarine is powered by conventional diesel electric power and is much smaller than either the Charlie or the Echo II. Displacing 2,500 tons, it measures 280 by 31 feet and has a speed of 16 knots and a range of 15,000 miles unrefueled. Its weapons consist of four SS-N-3 surface-to-surface missiles and eight (perhaps ten) torpedo tubes. It, too, must surface to launch its missiles.

A word must be said about the necessity of these submarines to surface before they can fire their missiles because this is a characteristic that may limit their effectiveness. Once it surfaces the submarine must elevate its launching platforms and open its storage canisters before it can actually fire the missiles. Even though hydraulic mechanisms can accomplish this in a matter of minutes, it is an observable sequence that can be monitored by a surface ship or an aircraft patrolling overhead. This may give the other side some warning time, possibly, depending on the rules of engagement, time to make a preemptive attack. Moreover, when the missile is fired the canister belches smoke and the missile leaves a visible trail. The SS-N-3 missile is subsonic, which makes it vulnerable to attack by a fast interceptor aircraft like the F-4 Phantom or F-14 Tomcat. Both of these aircraft are designed to have this capability. Another factor is that long-range antishipping missiles require mid-course guidance.

The second type of submarine, the antisubmarine submarine that is frequently seen in the Mediterranean, is the Victor class, a nuclear-powered vessel that can reach speeds in excess of 30 knots. *Jane's* says, "its much increased speed makes it a menace to all but the fastest ships." [13] The primary mission of this submarine in the Mediterranean is to help locate and attack American Polaris/Poseidon submarines. It has eight torpedo tubes and is presumed to carry some torpedoes with nuclear warheads. This class of submarine could also be used to attack surface shipping with conventional or nuclear weapons. However, its speed and design have led Western experts to conclude that it would be inefficient for this purpose, a moving van doing a grocery shopping cart's job.

Amphibious Ships and Auxiliaries. The Soviet Mediterranean squadron also has a limited amphibious assault capability. A small contingent of Soviet Naval Infantry, as the Russian marine corps is officially called, and an assortment of PT-76 amphibious tanks and other heavy equipment are embarked aboard Alligator- and Polnocny-class landing ships. The Alligator displaces 5,800 tons, is 374 feet by 51 feet, and can carry a cargo load of 1,700 tons. The smaller Polnocny, which displaces 1,000 tons and measures 246 by 29 feet, can carry up to six tanks. The naval infantry has not been used in a real combat situation in the Mediterranean, though it has conducted training exercises on Egypt's northern coast.

The other ships of the squadron are support vessels, a mixture of oilers, maintenance, and auxiliary ships that rendezvous with warships at harbors or at anchorages to refuel and replenish supplies. The most interesting auxiliaries are the AGIs—for auxiliary general intelligence. To the ordinary eye most of these could pass for small coastal cargo ships or fishing trawlers, except for the array of antennae and other electronic sensors fitted in the rigging and superstructure.

Aircraft Carriers. The addition of two aircraft carriers to the Soviet fleet in the near future—carriers that will make their maiden cruises in the Mediterranean—leads us to examine both the technical characteristics of the ships and the impact on the naval balance.

The *Kiev*, the first of two Kiev-class aircraft carriers, was nearing completion at Nikolayev in the spring of 1975, while her sister ship, the *Minsk*, was still under construction at the same yard. These two ships, writes John Moore, editor of *Jane's Fighting Ships*,

> mark an impressive and logical advance by the Soviet Navy. The arrival of these ships has been heralded by Admiral Gorshkov's support for embarked tactical air as a necessity

[13] Ibid., p. 542.

for navies employed in extending political influence abroad, and by a softening of previous Soviet criticisms of this class of ship.[14]

Its basic dimensions are 925 feet long and 200 feet wide overall if the extreme width of the angled flight deck is included, though the hull beam is probably about 125 feet. The flight deck is 600 feet long and the ship displaces 40,000 tons. Its built-in weapons package is thought to be quite comprehensive, consisting of 1 quad battery of SS-N-10 surface-to-surface missiles, 2 twin SA-N-3 launchers (with a slant range of 20 miles), possibly 3 or 4 SA-N-4 launchers, 1 twin ASW launcher, 14 twin 57 mm. gun positions, and 2 multi-barrel, that is, 12-barrel antisubmarine rocket launchers. The presence of the A/S rockets suggests a hull-mounted sonar and/or VDS.

This ship will carry about 25 ASW helicopters, probably the KA-25 Hormone. The absence of steam catapults, arresting gear, and other pieces of equipment that are standard on existing Western-designed aircraft carriers leads analysts to conclude that the only type of fixed-wing plane that could operate from its deck is a vertical short takeoff and landing (V/STOL) aircraft. The Soviet Union has been experimenting with such a plane, the Freehand, a follow-on to the YAK-36 which first appeared in public in 1967. The YAK-36 is a twin-jet delta-wing plane and is thought to be subsonic, but little else about it is publicly known. Models of this aircraft have been tested at sea from a specially fitted pad on the *Moskva*, but as far as is known is not in full production. The only operational fixed-wing V/STOL aircraft in the world is thought to be the British-built Hawker-Siddeley Harrier, some models of which are used by the U.S. Marines as ground attack and close support aircraft and are expected to be used aboard the new Tarawa amphibious assault ships.[15] The enormous power required to vertically launch the V/STOL aircraft consumes so much fuel that its combat range is severely limited. This flaw can be partially overcome by in-flight refueling or by a short takeoff run where the aircraft will develop at least some lift with its wings to reduce the jet power required and thus save some fuel. The length and angle of the *Kiev*'s flight deck suggests that Soviet engineers are aware of this solution to the V/STOL problem.

---

[14] Ibid., p. 532.

[15] The U.S. Navy is also developing the V/STOL plane, the XFV-12A, a single-seater aircraft for shipboard use. Engineers hope it will be capable of speeds in the Mach 2 range, or twice the speed of sound.

Assuming that the *Kiev* has 25 such aircraft with a combat range of, say, 500 miles, and an assortment of guns, bombs, and rockets, perhaps a 5,000- to 8,000-ton weapon load (like the Harrier), the U.S. Sixth Fleet and the West will have lost their monopoly on sea-based air power.

There are several factors, however, that will prevent the Soviet navy from achieving parity at sea in the near future even after their new naval air power is operational. First, unless there are radical breakthroughs in design and engineering, V/STOL aircraft will not be a match in air-to-air combat for the F-4 or F-14 or the older generation of Western carrier-based fighters. To date, Soviet V/STOL pilots lack the speed, maneuverability, and range to be a match for combat-seasoned American pilots in better planes. This will restrict the role of these new ships primarily to ASW.

Second, just in terms of numbers, two aircraft carriers are no match for the United States, which could in a crisis send three or more to the Mediterranean. In numbers of aircraft and firepower, the *Kiev* and the *Minsk* would be overwhelmed. It is likely that the Soviet Union may build more ships in this class, but they would not be available until well into the future, perhaps not until the late 1980s.

Last, in only a few months or even a year or two, the Soviet navy cannot gain the experience nor achieve the level of competence in the use of aircraft carriers that the U.S. Navy has acquired in more than 55 years and four wars. The aircraft carrier is perhaps the most complicated weapons system in the world. There will inevitably be some mistakes, some hit-and-miss operations, and a considerable lapse of time before Soviet sea-based air power earns tactical credibility.

More important than the marginal tactical advantage that the deployment of the *Kiev* will achieve is the profound psychological and political impact the ship will have on the littoral countries of the Mediterranean—or the countries of any other seas where this class of ship may cruise. Perhaps this is impossible to assess completely until after the fact. But the principal talking point of the U.S. Navy when comparing its strength with the Soviet navy has been the absence of Russian sea-based air power, which will disappear when the *Kiev* makes her debut.

Other Considerations. In addition to the characteristics of naval hardware, several other considerations must be taken into account in an assessment of the Soviet presence in the Mediterranean. Nearly all of them place limits on or otherwise constrain the Soviet Union's ability to make full military or political use of the Fifth Escadra. One aim of the Soviet policy in the Mediterranean is to neutralize these problems.

Lack of naval bases. The first limiting factor is the lack of an extensive and sophisticated network of bases and other facilities, which puts the Fifth Escadra at a substantial disadvantage compared to the Sixth Fleet. The Soviet navy in the Mediterranean, therefore, does not have the depth or security of lines of supply that would enable it to operate in strength or for very long throughout the sea.

This is one reason why the Fifth Escadra is concentrated in the eastern basin which is closer to Russian naval bases in the Black Sea and to the few naval facilities the U.S.S.R. has in the sea.[16] The Fifth Escadra regularly uses facilities in the Egyptian ports of Alexandria and Mersa Matruh or anchors at Sollum, though, since July 1972, the use of Mersa Matruh has dropped off slightly. Under the supervision of Russian technicians, this previously small harbor was deepened and improved to handle large vessels, and extensive storage and ship-repair facilities were constructed. Soviet ships frequently use the Gulf (or bay) of Sollum, about 100 miles to the east of Mersa Matruh, as an anchorage.

In Syria, there are usually Russian warships in Latakia, Banyas, and Tartus. There are areas in each of these harbors that are set aside for the use of the Soviet navy. Soviet submarines at times call for repair work at the Yugoslav port of Tivat, located in a fjord-like inlet just north of the Albanian border, which is a major ship-building and maintenance complex.[17]

There are conflicting reports about the Soviet naval presence in Algeria. One source on world arms and force deployments, the Stockholm International Peace Research Institute (SIPRI), reports both the existence of Soviet refueling stations at Algiers and Oran and of facilities for taking on potable water and recharging batteries at the former French naval base at Mer El Kebir, as well as the official Soviet denial of their existence. Algeria's armed forces use Soviet equipment almost exclusively, except for some light tanks supplied

---

[16] One could also reason that the bulk of the Soviet navy is in the eastern Mediterranean because the bulk of Russia's foreign policy investment is there and this would be the site of any conflict it would be likely to engage in.

[17] In an interview with the *New York Times*, Egyptian President Sadat said that the Fifth Escadra does not have "bases" but rather depots for reserve stocks, storage, and spare parts and not permanent shore installations. He went on to say that Egypt may make naval facilities available to all nations on the model of Yugoslavia, where warships may call for repair but the work is done by the country's nationals. *New York Times*, 22 April 1974.

Table 6

SOVIET NAVY ANCHORAGES, 1974

West of Melilla[a]	Gulf of Hammamet[a]
Lampedusa Island	Hurd Bank
Cape Passero	Kithira[a]
Gavidhos Island[a]	Gulf of Sollum[a]
Limnos Island	East of Crete[a]
Ras Al Kanais	Cape Andreas
Alboran Island	South of Crete

[a] Locations Soviet ships tend to favor.
Source: U.S. Navy Office of Information.

by France. However, President Boumedienne's strongly nationalisti government professes a nonaligned policy, which raises serious ques tions about the existence of naval and air base rights in Algeria Soviet warships do call at Algerian ports.

Anchorages. To help compensate for the lack of regular acces to ports in the western and northern Mediterranean, Soviet warship make use of shoals, areas where the sea is shallow enough in inter national waters for ocean-going vessels to anchor. Since these shal lows are located in or near choke points, they are ideal positions fo intelligence-gathering ships (AGIs) and vessels with enough spee and sea-keeping ability to conduct tattletale missions.

The bottoms of the shallows vary greatly so that some of th locations hold anchors better than others in rough weather. A list o the anchorages the Soviet navy used in 1974 is given in Table 6 Not all of these were used simultaneously, however.

These anchorages are also used to give the ships and crews a res and the opportunity to take on fuel and other supplies. Because th Soviet navy has not perfected the technique of underway replenish ment, they must make frequent port visits or rendezvous with oiler and supply ships at anchorages. While cruising as a guest on th U.S.S. *Springfield* in 1972, then the Sixth Fleet flagship, this write observed three Soviet Kotlin-class destroyers, an AGI, and an oile on 10 September at anchor off the coast of Kithira Island at th western entrance of the Aegean Sea, an important choke point.

These anchorages, however, only partially compensate for th absence of a base system. By contrast, the United States and Britain two non-Mediterranean powers, regularly conduct airborne maritim surveillance missions from several air bases at different locations ir

Table 7

FIFTH ESCADRA VISITS IN THE MEDITERRANEAN,
1964 AND 1974

1964	1974 [a]
Yugoslavia	Algeria
	Egypt
	France
	Morocco
	Syria
	Tunisia
	Yugoslavia

[a] Also during 1973 Soviet warships called at the Italian ports of Messina and Taranto, the first year such visits had occurred.
Source: U.S. Navy Office of Information.

the basin. These flights are made not so much to keep track of Soviet surface ships, because this can be done with high-frequency direction finding (HF/DF or "huff-duff") from remote sites [18] or satellites, but as part of an ASW effort. On balance the Soviet Union is more keenly interested in ASW in the Mediterranean than the United States because of the presence of Polaris/Poseidon SSBNs targeted against Russia. But even with the new aircraft carriers, the Soviet Union will not have the same ability to stage regular, long-range airborne ASW missions in the Mediterranean. For a while, the U.S.S.R. had exclusive use of six airfields in Egypt, some of which were used for bases for twin-jet TU-16 Badger aircraft that did fly such patrols. These flights stopped in July 1972 when President Sadat told Moscow to withdraw its troops from Egypt. The Soviet Union has long-range maritime surveillance aircraft capable of taking off from Soviet territory, overflying the Mediterranean, and returning to the same base; however, such flights would require overflight clearance from a littoral country.

Even the best anchorages can be untenable in a violent winter storm in the Mediterranean, particularly for the light displacement AGIs and destroyers that may be positioned for electronic eavesdropping or tattletale missions. In the absence of land-based facilities or protected harbors from which to stage such operations, Soviet ships must spend a lot of uncomfortable time at sea.

[18] The U.S.S.R. uses the same method for tracking surface ships.

Role of merchant shipping. The role that Soviet merchant shi ping plays must also be included in the equation. Non-naval shi and fishing fleets, particularly tankers, are frequently used as 1 plenishment vessels. American naval experts also say that Russi, merchant ships, cargo, fishing, as well as some passenger vesse have an intelligence-gathering capability and that their crews reg larly report on Western naval activities. In the shipping busine little, if any, political importance is attached to a Soviet merchantm; visiting a port to unload or take on cargo. And if that ship later mee a SAM Kotlin at, say, the Kithira anchorage and transfers food, wate or fuel, no custom or law has been violated.[19]

Montreux Convention. Another factor that limits the flexibili of the Fifth Escadra is the constraint that the terms of the Montre Convention put on Soviet warships using the Turkish Straits. (T text of the Convention appears in Appendix B.) This is a very re and at times awkward disadvantage for the Soviet Union. Officiall the Fifth Escadra is part of the Black Sea Fleet [20] and the Montre Convention must be complied with for the fleet to deploy or reinfor itself. By comparison, ships assigned to the Sixth Fleet enter t Mediterranean and leave through the international Strait of Gibralt under the rule of unrestricted and unannounced passage.

Ironically, the Montreux Convention, concluded on 20 July 193 was welcomed by the Soviet Union as an effective measure to preve foreign powers, specifically Germany, from sending large naval forc into the Black Sea.[21] The same terms of the convention that Mosco

---

[19] This factor should not be underestimated. The Soviet Union's merchant fle ranks first in the world in number of ships, with 7,123 vessels of 17,396,9 gross tons. See *Jane's Fighting Ships 1974-75* quoting the *Lloyd's Register Shipping*, p. 530. In carrying capacity the Soviet Union is tied in sixth place wi ships registered under the Panamanian flag of convenience. The statistic for t United States is 4,063 vessels of 14,912,432 gross tons, which ranks it eighth total number of merchant vessels and tenth in carrying capacity (*Jane's Fighti Ships 1974-75*, p. 370). American merchant ships are not used in the same w; in the Mediterranean.

[20] For all practical purposes the Fifth Escadra is a fleet in size, fire power, a influence. The reason that the Soviet Union has not officially called it a fle is probably political, for it could then be said that Russia had a fleet in a s not washing its own shores.

[21] For full and comprehensive discussions of the Turkish Straits and the evol tion of treaties regulating them, see Vali, *The Turkish Straits and NATO*, Har Howard, *Turkey, The Straits and U.S. Policy* (Baltimore and Washington: T Johns Hopkins Press in cooperation with the Middle East Institute, 1974), a Robert G. Weinland, *Soviet Transits of the Turkish Straits: 1945-1970—, Historical Note on the Establishment and Dimensions of the Soviet Nav Presence in the Mediterranean*, Professional Paper No. 94 (Arlington, V; Center for Naval Analyses, 1972).

endorsed as enhancing its security are now viewed as a nuisance or obstacle, though this theme has been muted since Stalin's death in 1953.

Broadly, the pact requires that Black Sea powers, including the Soviet Union, Bulgaria, and Rumania, give at least eight days' notice of their intention to send warships through the Turkish Straits. The name, type, and hull number of the vessel as well as the date of intended passage must be announced. There is a five-day grace period during which the ship may transit without filing a new notice. Each ship must transit escorted by no more than two destroyers and must complete its voyage in daylight.[22] At any one time there cannot be more than nine warships in the waterway and their aggregate displacement cannot exceed 15,000 tons. This means that only a Kara-class cruiser and a Krivak destroyer, whose aggregate displacement is 14,700 tons, could make the trip in one day. An exception is made for what the convention calls "capital ships" whose displacement is in excess of 15,000 tons.

Though its language is not specific, the convention appears to exclude aircraft carriers from transiting the straits, though such ships may enter at the invitation of Turkey for visits to ports within the straits.[23] To get around this provision, the Soviet Union calls the new angled-deck carrier *Kiev* and the helicopter carriers *Moskva* and *Leningrad* antisubmarine cruisers, a practice that Turkey accepts.[24] Soviet submarines rarely deploy to the Mediterranean from the Black Sea. They can, under the convention, transit the straits singly and on the surface but their declared destination, given on the notification to pass, must be to shipyards for repairs.[25]

Another provision says that if Turkey feels threatened by imminent danger of war, or if Turkey is a belligerent in a war, it has the right to prevent passage of warships through the straits. If a simple statement of closure does not prevent warships from forcing the straits, the waterway can easily be sealed with mines.

There are several practical effects of these provisions. The first is that because of its special relationship with Turkey, the United States is aware of the announced intention of Soviet warships to enter the Mediterranean through the straits. This gives the Sixth Fleet some time to make any adjustments in its own deployment. There could be several reactions. If notification is given that the *Moskva*

---

[22] Montreux Convention, Articles 11, 12, 13, and 14, in appendix.
[23] See Vali, *The Turkish Straits and NATO*, p. 46.
[24] Moore, *Jane's Fighting Ships 1974-75*, p. 532.
[25] Montreux Convention, Article 12, in appendix.

71

will transit the straits eight days later, a tattletale mission may b
laid on using a ship with special sensing devices to monitor th
helicopter carrier's operations. It may also be decided to alter th
patrolling pattern of a Polaris/Poseidon missile submarine to avoi
giving the ASW ship the opportunity of making contact.

To compensate for this the Soviet Union often attempts to spo
Sixth Fleet planning by filing multiple advance notifications whic
create some confusion about the actual date of transit. In realit
U.S. naval intelligence monitors Soviet naval movements in the Blac
Sea and will probably have some warning before a warship nose
into the straits. And a destroyer can always be assigned to a tattle
tale mission on relatively short notice.[26] The practice of multipl
advance notifications also permits the Soviet Union to partially ove
come the handicap of having to inform Turkey of intended passag
and gives it some flexibility to respond to a crisis. In a crisis, th
procedure would be something like this: as tension began to rise th
Soviet Union would make bookings for an assortment of ships b
type, hull number, and tonnage. After eight days, at the rate of tw
or three ships per day, the Fifth Escadra could be reinforced from th
Black Sea at a rather quick pace. In the same amount of time, how
ever, the United States, assuming it had additional naval forces avail
able in the Atlantic or Indian Ocean, could reinforce the Sixth Flee
through Gibraltar or the Suez Canal.

Black Sea Fleet. The ability to reinforce is hampered by anothe
factor, the size of the Black Sea Fleet. In a war scenario, it must b
assumed that the Black Sea Fleet would have a defensive mission an
if too many of its ships rushed to the Mediterranean during a regiona
crisis it could be caught without sufficient forces and unable t
recover them since the Montreux Convention applies to warship
entering as well as leaving the Black Sea.[27]

Political Uses of the Fifth Escadra

Port Visits. Despite all of these obstacles, the Soviet Union maintain
a formidable naval force in the Mediterranean and its presence ha
gained wide acceptance. Let us look at the Fifth Escadra's port visits
Soviet naval port calls in the Mediterranean differ from those of th
Sixth Fleet. First, fewer countries are involved, though the list o

[26] The distance from Odessa to the Bosporus: 375 miles, trip of 18.5 hours fo
ships making 20 knots.
[27] Montreux Convention, in appendix.

ports has dramatically expanded since 1964, as Table 7 indicates. While the visits to Italy, first made in 1973 and not repeated in 1974, have been infrequent, they are not without significance.[28]

The second major difference is one of style. The men are more rigidly controlled. Soviet sailors are not allowed to go ashore individually. They wear their uniforms, walk in groups of five or six, and are always accompanied by a petty officer. They do not frequent bars, and the times and places for souvenir shopping are arranged. The sailors rarely eat more than one meal in the local restaurants during the port visit.

Soviet ships in the Mediterranean tend to spend more time in port or at anchorages—about 60 percent—than time underway. When the ships do appear to be conducting training exercises, normally only a few vessels take part and the maneuvers rarely last more than a few days. There have, of course, been exceptions.

The Libyan Coup d'Etat. One of the more intriguing activities of the Soviet Fifth Escadra took place between late August and mid-September 1969 at the time of the 1 September coup in Libya that brought Colonel Qaddafi to power. Earlier in the year, naval officers from the Soviet Union, Egypt, and Syria had planned an operation to practice an amphibious landing on the Egyptian coast 20 miles southwest of Alexandria. While a joint force of marines stormed the beaches in a mock assault, Soviet, Egyptian, and Syrian warships spread out in a 210-mile protective screen from the Gulf of Sollum to eastern Crete. (The Gulf of Sollum is on the Libyan border and just 70 miles east of Tobruk and Al Adem where, at the time of the coup, Britain maintained tank bases.)

This massive force of nearly 100 ships from three countries included more Soviet naval vessels than had ever assembled before in the Mediterranean—70 surface ships and submarines plus the helicopter carrier *Moskva*. They were armed with an impressive array of guided missiles, both surface-to-surface and surface-to-air, radar-directed guns, antisubmarine weapons, and electronic warfare equipment—almost every tactical naval weapon in the Soviet inventory.

The exercise is interesting for several reasons. It gives an indication of the capability of the Soviet navy, indicates some future uses

---

[28] To put port calls into proper perspective, it must be understood that the frequent use of Tartus, Syria, by the Fifth Escadra commander is not equivalent to an occasional visit by one or two Soviet warships to Tunis. Sixth Fleet units also visit Tunisian ports, but Sixth Fleet ships do not visit Syria at all.

SOVIET NAVAL EXERCISE, 1969

MEDITERRANEAN SEA

ITALY
SICILY
ROME
Naples
TRIPOLI
Misratah
Marsa al Burayqah
Az Zuwaytinah
BENGHAZI
Al Bayda
Darnah
Al Adem
Tobruk
Sollum
LIBYA
Marsh
EGYPT

YUGOSLAVIA
ALBANIA
TIRANE
Durres
Vlore
GREECE
ATHENS
Piraeus
MALTA
VALLETTA

BULGARIA
SOFIA

TURKEY
ANKARA
Istanbul
Eceabat
Izmir
Mersin

CRETE
Iraklion

CYPRUS
NICOSIA
Famagusta

SYRIA
DAMASCUS
Latakia
Tripoli
LEBANON
BEIRUT
Haifa
ISRAEL
TEL AVIV–YAFO
Ashqelon
Port Said
Suez
Rosetta
Alexandria
CAIRO
Maruh
Sollum

IRAQ

JORDAN
AMMAN
Elat Al Aqaba
GULF OF SUEZ

SAUDI ARABIA

(Likely Flight Pattern of U.K. Transports)

(Joint Amphibious Force)

NOTE: The three U.S. Sixth Fleet aircraft carriers went, of course, escorted by cruisers, destroyers, and submarines.

LEGEND:

U.S. Aircraft Carriers

U.S. Surface Ships

(Hypothetical Western Task Force)

Soviet Surface Ships

Soviet Missile firing Submarines

0   100   200   300 Statute Miles
0   100   200   300 Kilometers

74

of Soviet naval power in the Mediterranean, and suggests that the Soviet fleet may have been used by Egypt and the officers around Qaddafi as a "floating trip-wire" for the Libyan coup d'etat.

A detailed analysis of the joint operation was made by American intelligence experts. This was based on the movements of the Soviet fleet, particularly the positioning of their missile ships, their radio and radar transmissions, and other information from a wide variety of sources. These data revealed that the Soviet role was to neutralize a hypothetical Sixth Fleet carrier task force that was moving from the western Mediterranean to attack the amphibious operation taking place behind the screen. During the exercise Soviet forces in the screen relied heavily on the use of surface-to-surface and surface-to-air guided missile ships to lead the imaginary counterattack. The judgment of the experts was that the Soviet force would have the capability for a "strong" assault against attacking aircraft, surface ships, and submarines in a real situation.

At the same time the exercise demonstrated one way that the Soviet Union could assist an Arab country if a confrontation with Israel occurred or if a real American force attempted to intervene in a local flare-up such as Lebanon in 1958 or Jordan in 1970. What is not known, though, is at what point the Soviet Union would give the order to open fire. This is, of course, a political question and one that ought to be the object of constant attention.

This brings us to the cost-absorbing aspect of the exercise: what relationship did the naval maneuvers so close to Libya have to the timing of the Qaddafi coup? There is nothing from public sources to suggest that the exercise was timed to coincide with the young officers' bid for power. However, there are strong indications that it might have happened the other way around: the coup might have been timed to coincide with the exercise. The Libyan officers were in touch with the Egyptians before the coup, and the Egyptians were involved in the planning of the naval exercise. It is conceivable that the idea could have surfaced that late August or early September would be the best time for a coup attempt because of the presence of the Russian fleet, whose role in the exercise might deter any countermove by the United States and Britain.[29]

Some information can be gleaned from the public record. On 18 August 1969, the *Moskva* passed through the Turkish Straits

---

[29] The Egyptian newspaper *Al Ahram*, for example, reported that the young officers had planned a coup attempt earlier in the year and that twice—24 March and 5 June—they had postponed their move. Quoted in the *New York Times*, 6 September 1969.

escorted by two guided missile destroyers, topping off an unprecedented Soviet naval build-up in the Mediterranean. Two days later, the Pentagon announced that Soviet naval forces away from home waters had reached a record size. Department of Defense spokesman Daniel Z. Henkin said that of the 125 ships deployed abroad, more than half were in the Mediterranean.[30] Henkin added that even though the Soviet forces outnumbered the Sixth Fleet, the U.S. Navy in the Mediterranean was in a "high state of readiness" and that the fleet's two aircraft carriers made American forces "far more powerful." He went on to say that the Russian helicopters and short-range missiles were limited to a range of several hundred miles.

Then, 48 hours before the Libyan coup, John Cooley, a well-informed writer for the *Christian Science Monitor*, touched closer to what was to unfold. On 30 August 1969, he reported from Beirut, "The Soviet move has raised Arab speculation that Moscow is trying to *lift the morale* of its Arab friends in their confrontation with Israel."[31] Tripoli Radio, the broadcasting station of the newly installed Libyan regime, announced on 4 September that the presence of Soviet warships had deterred British intervention against the coup. In his book *Gunboat Diplomacy*, British diplomat James Cable makes this observation: "Although it is most unlikely that British intervention was ever contemplated, there were in fact Soviet warships in the vicinity and this was well known in London as well as Libya. What mattered, however . . . in Libya . . . was what people believed."[32]

This belief was subtly encouraged throughout the Arab world by Radio Cairo. On 26 September during a panel discussion it broadcast on the role of the Soviet fleet in the Mediterranean, one commentator said, "The presence of the Soviet fleet at this strength is a guarantee for neutralizing the effectiveness of the Sixth Fleet and deterring it from carrying out new imperialist adventures."

The general Arab view that the United States and Britain seriously considered—but then rejected—intervention was strengthened by several factors. First, the deposed Libyan monarch, King Idris, who was vacationing at a health spa in Turkey when the coup occurred, asked the United States and Britain to help. These appeals were later reported in the press and received wide coverage in the

---

[30] *New York Times*, 21 August 1969.

[31] Emphasis added. The day after the coup, without mentioning events in Libya, Peking Radio branded the Soviet exercise "an outrageous demonstration of . . . social-imperialist 'gunboat policy.' "

[32] James Cable, *Gunboat Diplomacy: Political Applications of Limited Naval Force* (London: Chatto and Windus for the Institute for Strategic Studies, 1971), p. 147.

Middle East. Second, both the United States and Britain had military bases in Libya. The British armored forces stationed at Tobruk, where there was also an airfield, and at Al Adem were the most worrisome to the Libyan officers. The British kept the tanks ready to roll, only having to fly in crews from Cyprus. British contingency plans called for their possible use in a local dispute to repel an external ground attack from the east. The last fact that encouraged this interpretation was, of course, that the United States and Britain did not intervene.

On 3 September, the new government told the commanders of the U.S. bases at Wheelus and Tobruk to cease all training flights. "Although this was a precautionary move," the *New York Times* reported from Beirut, "there had been some speculation in Arab capitals that the treaty of friendship between Britain and the deposed Libyan regime, which provides for protection against 'internal subversion,' might be invoked." [33] This particular line of speculation evaporated after a widely publicized meeting in London between Foreign Secretary Michael Stewart and Omar Shelhi, an advisor to the king. Obviously embarrassed by reports of the meeting, the Foreign Office was moved to give a background press briefing saying that it had made it clear to Shelhi that the British government had no intention of intervening. [34] The British government was also aware that transports carrying the tank crews would have to overfly Soviet SAM guided missile ships to reach Tobruk.

The American planes at Wheelus, the field used to receive aircraft based in Europe for gunnery and bombing practice in the Libyan desert and as a way station for transiting aircraft, were less suited for employment in a domestic crisis. However, the overriding factor against Western intervention was that the new government was in charge of the country and it was not worth trying to alter a fait accompli at the risk of further worsening the Western position in the Mediterranean. The coup was almost immediately successful, with very little bloodshed. The new regime quickly announced that it would respect existing treaties, protect the lives and property of foreigners, and not interfere with the operation of foreign oil companies. The deposed king was out of the country and no earnest domestic support for him was manifested. The success of the coup was acknowledged the same day by Egypt and Iraq and by 4 September by Algeria, Syria, Sudan, and South Yemen. The Soviet Union extended formal diplomatic recognition on 5 September, and

[33] *New York Times*, 5 September 1969.
[34] *New York Times*, 6 September 1969.

after close consultation the United States, Britain, France, and Italy followed on 7 September.

Moroccan Troops to Syria. Another interesting Soviet naval operation in the Mediterranean occurred in early 1973 when Soviet warships helped the Arabs against Israel without taking a substantial risk of getting directly involved in any fighting. When King Hassan offered to send Moroccan troops to Syria to join in the planned attack on Israel, the U.S.S.R. transported these forces and their equipment from Algeria.[35]

The Moroccan troops arrived in Algeria by truck and then went aboard two warships for the trip. Chomeau says that it marked "the first time that foreign military units had thus been transported by the Soviet Union. Such a use of naval force was done at little risk to the U.S.S.R., but the potential of winning political favor with the Arab governments by such an act was great."[36]

The Outlook

Though Soviet diplomacy is conducted in great secrecy, there is enough that is visible to suggest that Russia is very much in the market for additional naval facilities in the Mediterranean and wants to hold on to and expand the installations it now has. In this context the Soviet Union successfully negotiated with Egypt for permission to join the Suez Canal clearing operation. We have already mentioned the publicity given to the participation of the U.S. Navy, which was viewed as a possible turning point in Egyptian-American relations. The Soviet navy did not want to be left out. Another likely reason for Russia's participation was that along with the mines and explosives, the bottom of the canal was strewn with parts of tanks, planes, and other weapons that Egypt had obtained from the Soviet Union and that Moscow wanted to keep out of American and British hands.[37]

Another indicator of Soviet concern about the Egyptian-American rapprochement was the visit of the Fifth Escadra commander, Rear Admiral Vladimir Ilich Akimov, to Alexandria between 12 and 17 December 1974, during the port call of the two Soviet guided missile ships, the *Groznyy* and the *Krasnyy Kavkaz*. The following week, the commander in chief of the Soviet navy, Admiral of the

---

[35] Chomeau, "Sea Power as a Political Instrument: The Soviet Navy in the Mediterranean," pp. 126-27.

[36] Ibid., p. 127.

[37] *New York Times*, 27 March 1974.

Fleet Sergei Gorshkov, visited Cairo. Though the details of that visit are not publicly known, U.S. government analysts believe that Gorshkov was trying to protect Soviet naval assets in Egypt.

The Soviet Union is also known to have asked Syria for permission to widen its port rights and possibly to stage maritime surveillance flights from Syrian airfields. Though Russian pilots have flown occasionally the new MiG-23 from these bases, the flights have been relatively short and not in a pattern over the sea that suggests reconnaissance missions.

Expansion of the naval facilities in Yugoslavia or a formal agreement to establish a Soviet naval base would obviously be desirable to Russia, but as long as President Tito remains in power it seems unlikely that the Soviet Union will be successful in obtaining such a base. What may happen in post-Tito Yugoslavia is the subject of much speculation and will be dealt with in the next chapter.

Deep natural harbors and extensive facilities exist in other parts of the Mediterranean where the Soviet Union has been trying to improve relations or already has forged strong links through its military assistance program. Shortly after Dom Mintoff became prime minister of Malta in 1971, the Soviet ambassador to London made a widely publicized visit to Malta in the midst of the talks to review the British base agreement. The visit caused alarm in London, Washington, and Rome and was a catalyst for the eventual conclusion of a new base agreement more or less on Mintoff's terms. (For a fuller discussion of this, see page 96.)

Libya, reportedly scheduled to buy $800-million worth of arms from Russia, would be a logical site for the Soviet Union to seek base rights for its navy and aircraft.[38] The press reports of the arms deal also said that the granting of base rights in exchange for making the military hardware available was part of the overall agreement. However, *Pravda* strenuously denied the reports of any plans to open bases in Libya. Such flurries of publicity about the policies of Libya have often turned out to be, if not entirely false, considerably inflated. Libya has been adamant about granting base rights to any foreign power since the ouster of King Idris in 1969. A decision to give maintenance and supply facilities to Soviet ships would be a major reversal of the Libyan policy goal of ending the superpower presence in the Mediterranean.[39]

---

[38] Ibid., 23 and 24 May 1975. Also see ibid., 21 February 1975, for reports of increased Soviet presence in Libya.

[39] The reports were first carried by Cairo's semi-official daily *Al Ahram* and mentioned a $4.4 billion arms deal. Following President Sadat's refusal to

Another ideal site is Bizerte in Tunisia which is a former French naval base and is now largely unused. Tunisia permits Soviet warships to visit, but has not granted anything approaching the rights extended in Syria or Egypt. Under President Habib Bourguiba, Tunisia has been consistently pro-Western and the West, principally France and the United States, supplied Tunisia's small armed forces. Tunisia too has called for the withdrawal of the fleets of both superpowers from the Mediterranean.

As the Fifth Escadra increases its complement of modern, long-range ships, it may occasionally operate in strength in the western Mediterranean, particularly after the *Kiev* enters service. But the absence of naval bases or an expanded list of available ports will hinder and perhaps even prevent any continuous presence in the western basin.

A U.S.-Soviet War Scenario. Now that we have weighed the military strengths and capabilities of the United States and the Soviet Union in the Mediterranean, we must ask how are they likely to behave if they fight each other? Any response to this question has to be speculative. However, the conduct of the Sixth Fleet and the Fifth Escadra during the 1973 Middle East crisis provides, in my judgment, a partial answer. As tension mounted between Washington and Moscow in the final days of that Arab-Israeli war, the eastern Mediterranean contained the highest concentration of American and Soviet warships in history. Both navies were poised to begin shooting at a moment's notice. Their actions illustrate what each expected the other to do and therefore make a good case study.

At 2:00 p.m. on 6 October 1973, Egypt and Syria launched a coordinated surprise attack on Israeli troops along two fronts. Initially the element of surprise worked to their advantage. The Egyptians overran the Bar Lev Line on the eastern bank of the Suez Canal and pushed into Sinai. The Syrians, along with the 1,800 Moroccan troops brought there in Soviet warships, were able briefly to dislodge the Israelis from their forward positions on the Golan Heights. However, by 10 October, Israeli forces had gained the initiative on the Syrian front and had started to divert reserve troops to Sinai.

---

merge Egypt with Libya in a single federal state, the two countries intermittently conducted intense propaganda compaigns against each other. However, Libya and the Soviet Union have significantly improved their relations since Libya's isolation in the Arab world. Soviet Premier Aleksei H. Kosygin ended a four-day visit to Libya on 15 May 1975. In a joint communiqué the two countries called for closing all foreign bases in the Mediterranean to create "a lake of peace." This at least raises the possibility that the *Al Ahram* report was calculated to embarrass both Russia and Libya.

Also on 10 October the U.S.S.R. began a massive airlift of supplies to Syria and Egypt. The planes, giant AN-22s and smaller AN-12s, obtained overflight permission from Yugoslavia for this operation. The United States began its airlift to Israel to replace planes, ammunition, and other hardware on 13 October. On 15 October Israeli forces crossed the Suez Canal. They continued to gain ground and were threatening both Cairo and the Egyptian Third Army that was trapped on the canal's east bank.

The United Nations Security Council passed its first cease-fire resolution on 22 October. This broke down almost immediately and Israel raced toward Suez City, pressing down on the Third Army. Another cease-fire resolution was adopted on 23 October. Then, in a surprise move, President Sadat called for a Security Council meeting and asked for the United States and the Soviet Union to send troops to the Middle East to enforce the tenuous cease-fire. Sadat was worried that the Israelis might decimate his Third Army, now at the mercy of the desert and Israel.

On 24 October Soviet Communist Party Secretary Brezhnev sent a note to President Nixon asking that both countries comply with Sadat's request and adding that if the United States would not send troops "we should be faced with the necessity urgently to consider the question of taking appropriate steps unilaterally." [40] In response to Brezhnev's note, on 25 October, Mr. Nixon ordered a Defense Condition Three (DEFCON III) alert of U.S. forces around the world. In the Soviet Union airborne units apparently destined for the Middle East had been put on alert at about the time the Soviet note was delivered.

After several hours of very intensive diplomacy the crisis subsided, with the Soviet Union indicating to the United States that it would not send troops to the Middle East and the United States virtually ordering Israel to spare the Third Army.

Meanwhile in the Mediterranean the Sixth Fleet and the Fifth Escadra were sending additional ships to the eastern basin. The United States sent another carrier task group, built around the U.S.S. *John F. Kennedy*, into the sea from the Atlantic. The Soviet Union had dispatched more warships, including 12 that passed through the Turkish Straits between 6 and 26 October. Both sides were careful to stay out of each other's way while monitoring each other's moves. During the U.S. airlift to Israel, the short-range A-4s flown by U.S. Navy pilots landed on carriers in leap-frog fashion along their route

[40] *Strategic Survey 1973* (London: International Institute for Strategic Studies, 1974), p. 47.

DEPLOYMENT OF U.S. SIXTH FLEET AND SOVIET FIFTH ESCADRA, 1 NOVEMBER 1973

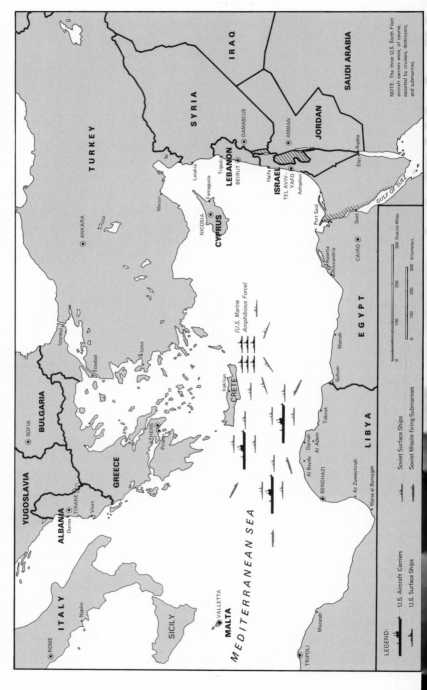

to refuel, and carrier-based fighters and early-warning aircraft gave protective cover for the A-4s and C-5A transports until the planes were within range of Israel's air defenses.

According to American analysts who have made a thorough study of the actions of both navies during this period, the Fifth Escadra commander had his ships in a preemptive position and was prepared to open fire on the carriers if carrier-based planes had been used to attack transports flying Soviet troops to Egypt. The position of Soviet ships in relation to Sixth Fleet units was this: Around each of the three American carriers were three Soviet ships. Two were destroyers, one of them a surface-to-surface missile ship, the other a surface-to-air missile ship. The third ship, with more conventional armament, was probably a tattletale capable of providing mid-course guidance for surface-to-surface missiles fired from another location. There were four Soviet missile-firing submarines on submerged patrol some distance away. Task Forces 61 and 62 were patrolling in relatively tight formation just south of Crete. They were shadowed by five Soviet warships, some of which carried surface-to-surface missiles.

To guard against this threat, the Sixth Fleet kept several scores of planes in the air watching closely for any sign that shooting was about to start. All of the fighter aircraft were armed. The complement of aircraft and their roles reveal a great deal about how the Sixth Fleet would have responded if shooting had started. The E-2 Hawkeye early-warning aircraft provided long-distance monitoring and strike control, while other planes equipped with electronic warfare gear were prepared to jam both the surface-to-surface and surface-to-air missile guidance systems of the Soviet ships. On the surface, the carrier defenses were manned in case any of the missiles were not intercepted by the fighters. Below the surface, American submarines were also ready to follow any contact and attack it.

Under what circumstances would shooting have actually started? It is impossible to say with any accuracy. The purpose of the American alert was to prevent the Soviet Union from sending troops to Egypt. Carrier-based fighters could have been assigned to shoot down the Soviet transports, though it is considered doubtful that this would have happened. The object for which the Soviet Union appeared to be ready to send in its troops was to prevent the destruction of the Third Army and a direct attack on Cairo. The massive build-up of naval forces on the scene and the high state of readiness of additional forces gave great urgency to the efforts to end the ground war by diplomatic means. Israel too knew of the naval build-up and other alerts and recognized the dangers of a Soviet-American conflict. This was emphasized by the United States during the diplomatic contacts that persuaded Israel not to press the attack on the Third Army and Cairo and to abide by the cease-fire. In short, as the danger expanded, the violence increased, and the possibility that nuclear weapons might be introduced became more apparent, diplomacy became the only acceptable alternative.

Probable outcome. The other intriguing question is: if shooting had started—and if the fighting had been limited to the two navies in the Mediterranean—how would it have turned out? [41] Even to attempt to definitively answer such a speculative question requires access to highly sensitive data that are not available to this writer—probably not to anyone. About Soviet capabilities one would have to know, for example, how much time it takes to elevate the surface-to-surface missile launchers and to eject the covers before actually firing the projectiles, how much time it takes to reload and fire a second missile salvo, and whether or not the missiles contain their own jamming equipment to guard against their destruction before reaching target. Obviously the technical manuals used by Russian sailors in the missile control centers are not readily available to the U.S. Navy. From the Americans one would have to find out the time it would take for the E-2 Hawkeye to detect a Soviet missile, determine its course, speed, and altitude, and pass the information on to a fighter aircraft. You would also need to know how much time it would take for the fighter's control system to digest the information, work out a solution for its own course and speed, and then attack the Soviet missile. Even if such data were available the reactions of the people who operate the systems could not be predicted.

Having said this, however, it is possible to make a few general observations with confidence. First, after the shooting had started it would have been the task of the Soviet ships to sink or otherwise incapacitate as many American aircraft carriers as possible in the shortest possible time. If the Soviet warships had not succeeded in the first 15 minutes, they probably would not have succeeded at all.[42]

---

[41] The assumption that any conflict between the Sixth Fleet and the Fifth Escadra would be confined to the Mediterranean and would not escalate into a global war is, of course, debatable. A commander of a ship on either side could in a moment of strain give the order to open fire. The opposite commander on the scene, who would have no way of knowing whether that first shot was the result of panic or an order from a higher command, would have to react by shooting back.

[42] In a general war, the assumption is that the Soviet "concept does not require the ability to survive a naval contest, since Russian ships would be within range

For within that 15 minutes Sixth Fleet aircraft and ships would have been attacking every Russian warship within range. The purpose of their strikes would not necessarily have been to sink the offending ships, but to cause enough damage to prevent them from firing their missiles and guns. This would have meant destroying the electronic gear in the masts and the missile and gun positions. The submarine threat is a different matter altogether, because the ASW effort could have continued for days—even weeks—given the conditions of the Mediterranean and the capabilities of submarines.

There has been much discussion of the combat experience of the U.S. Navy during the Vietnam War and speculation that this would give the United States an edge over the Soviet navy in any conflict. It is true that carrier-based pilots flew many missions over North Vietnam and encountered surface-to-air missiles and MiG-21s while Seventh Fleet ships operated in the Gulf of Tonkin and off the coasts of the two countries. However, the Seventh Fleet did not have to contend with any significant opposition unless it purposely got within range of coastal batteries or patrol boats that cruised near the shore. Neither did the Seventh Fleet have a surface-to-surface missile threat to deal with. In the Mediterranean the situation is different. A former Sixth Fleet commander, Vice Admiral Gerald E. Miller (Ret.), expressed concern about this difference and cautioned against what he called "The Gulf of Tonkin Syndrome" when planning scenarios in the Mediterranean. It would be fascinating to evaluate the chances of survival of two American aircraft carriers under concerted attack by a combination of missile-firing surface ships and submarines. A scenario could be constructed where 48 missiles would be launched simultaneously from Soviet surface ships,[43] while two or three Charlie-class nuclear-powered submarines attacked with both missiles and torpedoes. The outcome of such a conflict would be an open question.

So far we have been looking at only two elements in the Mediterranean power balance—the United States and the Soviet Union, prin-

---

of American units at the outset of war, but only the ability to survive long enough to launch their missiles at U.S. carriers. . . . In short, the role of the Soviet Mediterranean fleet in a general war involves a suicide mission." C. B. Joynt and O. M. Smolanski, *Soviet Naval Policy in the Mediterranean*, Research Monograph No. 3 (Bethlehem: Department of International Relations, Lehigh University, 1972), p. 15.

[43] A typical configuration could be: two Kresta II, 16 tubes; one Kara, 8 tubes; two Krivak, 8 tubes; and two Kynda, 16 tubes. The numbers of surface-to-surface and surface-to-air missile launchers and any reload capability are a standard part of the daily intelligence briefings of senior Sixth Fleet officers, as are the number and location of surface warships and submarines.

cipally in the setting of the Middle East. It is true that Soviet American rivalry in the Middle East is more important in sheer weigh of presence and energy expended than any comparable confrontatio in the region. But there are other elements of power and other dis putes outside the Middle East that could cause a crisis or alter th balance of power in the Mediterranean context. Let us now loo at these.

5
SCENARIOS FOR CRISIS AND CONFLICT

An Overview

When it comes to formulating scenarios for crises, assessing the relative strengths of armed forces in the Mediterranean, and determining how they affect American interests, it would be absurd to try to list the infinite number of possible permutations the political geography of the Mediterranean Basin could provide. For example, it is theoretically possible to construct a model that would have Algeria, Greece, and Egypt in concert declaring war on Italy and invading Sardinia. Or Lebanon could lay claim to Tunisia because the Phoenicians once ruled Carthage. Either of these developments would certainly cause problems for the United States. However, neither is likely to occur.

Also, in evaluating the armed forces of the Mediterranean region, different measuring systems must be used for different countries. The methods used to weigh the strengths and capabilities of the United States and the Soviet Union in the Mediterranean will not be applicable to Morocco and Spain, Greece and Turkey, Israel and Egypt. There are several major differences between the forces of the superpowers and those of Mediterranean countries. With few exceptions, armed forces in the Mediterranean are neither designed nor equipped to project their power beyond the immediate vicinity of their own territory. Most perform two basic missions: (1) preserving internal security by supplementing the efforts of urban and rural police forces keeping the domestic peace and, in some cases, maintaining the government in power; and (2) defending the territorial frontiers. This explains the heavy emphasis on infantry and armor, and the fact that most of the air forces consist of planes designed for ground support and air-to-air combat rather than long-range delivery of weapons. Again, with few exceptions most Mediterranean navies can only perform coastal defense missions; many are confined to antismuggling and sea rescue roles. This is not to say that there are no strategic issues at stake there are. A nation's very survival may depend on its ability to check firmly an insurgent group, or to deal decisively with armed cross border raids, or to launch a preemptive attack when a neighbor has assembled threatening troop formations, or to maintain an air defense system strong enough to discourage air strikes and reconnaissance sorties. Indeed, violence of strategic proportions is reached much more quickly in the Mediterranean than elsewhere. Because distances are shorter, there is a greater likelihood that urban centers could come under attack, subjecting the civilian population to heavy casual ties and leaving many homeless refugees; and there is a greater chance of involving one or both of the superpowers. The Arab Israeli wars and the 1974 Turkish invasion of Cyprus provide exam ples of this.

Another important factor is that many economic, social, and political variables in the Mediterranean combine to cause marked imbalances in military strengths. For this reason the industrial and mineral wealth, level of education, technical achievement, motivation religious and ethnic composition, and command and control structure between the political and military leadership in a given country become just as important as numbers of men, tanks, planes, and ships. This also means that few countries in the basin are capable of substantially helping maintain a balance of power favorable to U.S. interests. Outside the island countries of Malta and Cyprus there are more than 2 million army troops, 31,000 tanks and other armored vehicles, and about 3,200 combat aircraft around the Medi terranean. These aggregate figures are staggering and indicate a great capability for creating violence—either internal or international— rather than an ability to act in concert.

Even among the few countries that can exert a calming influence perhaps none will be willing to closely identify themselves with the United States during a crisis. It is one thing for Mediterranean coun tries to conduct joint training exercises with the Sixth Fleet using a hypothetical Warsaw Pact invasion of Thrace as the scenario—a circumstance clearly calling for a NATO response. It is quite another matter for the same countries to permit American planes to overfly their air space or to land for refueling during an emergency airlif to Israel—a concession that the economics and politics of oil just a clearly preclude and that NATO would not cover.

If we carry this line of thinking through, we come to the conclusion that because of deep political divisions (either apparent or just under the surface), economic dependency, or military weakness—or a combination of these—most Mediterranean countries contribute to the problems facing the United States rather than sharing in its efforts to preserve stability in the basin. Let us start, then, at the western entrance to the Mediterranean and assess the balance of power in the context of realistic scenarios and American interests.

Outstanding Disputes and Political-Military Problems *

As we enter the Mediterranean from the west, we encounter three disputes that could lead to armed confrontations jeopardizing the vital right of transit through the Strait of Gibraltar. The disputes concern the Spanish Sahara, the Spanish enclaves and islands along Morocco's Mediterranean coast, and Gibraltar itself. These disputes are related and all have implications for American policy far beyond the importance of the contested real estate. The reason is that the three main protagonists—Spain, Morocco, and Britain—are either allies of the United States or pro-American. Spain is on the threshold of the post-Franco era and is seeking closer security ties with the United States and Western Europe while asking a higher price for the U.S. bases on its territory. In Morocco there is the ever present possibility that political violence may topple the pro-Western monarchy of King Hassan and replace it with a Libyan-style Arab socialist regime. In 1971 and 1972 there were two organized attempts on Hassan's life, both of which were openly endorsed by Libya's Radio Tripoli. And Britain, while planning to keep a token force in Gibraltar for the foreseeable future, has announced severe cuts in its overall military posture in the Mediterranean which makes its presence in Gibraltar at least psychologically more vulnerable to a determined attempt by Spain to take the rock by force. Under the present Spanish government this may not occur, but after Franco a weak government badly needing domestic support might act precipitously.

Spanish Sahara. Spanish Sahara is the object of dispute among a number of protagonists: Spain, Morocco, Mauritania, and Algeria. Occupied by Spain since the nineteenth century, this territory has been claimed by Mauritania and especially by Morocco, which con-

---

* Editor's Note: Though this chapter was completed in June 1975 and may be somewhat overtaken by events, many of the basic problems remain and therefore continue to deserve attention.

siders all Spanish possessions in northwest Africa (including Mauritania) Moroccan territory. Morocco's claim is based on the fact that Moroccan rule prevailed throughout the area in medieval times and is spurred by the recent discovery there of the world's largest known phosphate deposits. Mauritania claims the territory because its citizens and the inhabitants of the southern Spanish Sahara province, Rio de Oro, are of the same racial stock.

After gaining its independence from France in 1956, Morocco sponsored agitation in the area by supporting local forces and deploying Moroccan irregular troops. However, Spanish and French pacification efforts in 1958 succeeded in restoring order on the Moroccan and Mauritanian border areas. During 1964 and 1965, Morocco and Mauritania took the question to the United Nations, which adopted a resolution calling on Spain to liberate the Spanish Sahara and begin negotiations concerning its destiny. A subsequent resolution in December 1967 required Spain to sponsor, in consultation with Morocco and Mauritania, a referendum in Spanish Sahara to determine its future. Spain accepted the principle of self-determination, but opposed the annexation or partition of Spanish Sahara by Morocco or Mauritania.

In 1970 and 1973, following the settlement of various disputes among their countries, the heads of state of Morocco, Mauritania, and Algeria jointly affirmed their desire to achieve the decolonization of Spanish Sahara.[1] The Algerian interest in this affair was based not on any territorial claims, but instead on a desire to effect the creation of an independent state in Spanish Sahara rather than the enlargement of either of its neighbors. A further United Nations resolution urging a referendum added momentum to the movement toward self-determination. Meanwhile, Spain began a program of "progressive participation" of Spanish Saharans in their own self-government—measures designed to result in internal autonomy and finally full self-determination.[2] The Moroccan reaction to this program was unfavorable, claiming that Spain was attempting to establish a puppet state in opposition to the real will of the people and warning Spain against taking such unilateral action. Spain responded by accusing Rabat of a "premeditated annexationist campaign" in disregard of the right of self-determination, by announcing a referendum to be held under United Nations auspices in 1975, and by allocating special funds for the general public welfare of the territory's native popula-

---

[1] *New York Times*, 14 September 1970.
[2] Ibid., 28 September 1973.

90

tion.[3] On 23 May 1975 Spain announced that it was ready to grant independence to Spanish Sahara "in the shortest period possible," though no timetable was set.[4] The Spanish move appeared to be motivated by an increase in guerrilla activity and a wish to reduce the number of international problems facing the Spanish government that will succeed Franco.

Relations among the various participants in this dispute are not harmonious. Despite the solidarity of Spanish Sahara's neighbors on some issues, their respective interests over the future of Spanish Sahara sharply diverge. Mauritania has launched its own diplomatic campaign to muster support for its claims and has thereby provoked the Moroccan charge of troublemaking. Spanish Saharan exiles have banded into liberation organizations in the neighboring countries. They seek the support of these governments and claim responsibility for numerous disturbances.

Spain has sufficient military muscle on the scene to deter any move by regular Moroccan or Mauritanian forces against Spanish Sahara. Ten thousand well equipped Spanish troops are permanently stationed there, including two crack Spanish Foreign Legion regiments. Spanish air force planes are stationed at several airfields and others could be flown from the nearby Canary Islands or from the mainland. Spain also has a significant amphibious capability and a large enough army (208,000) to enable it to reinforce the troops already there. Even if Spain did not possess such overwhelming local superiority, it is unlikely that either Morocco or Mauritania would attempt such a set piece confrontation in a territory nearly as large as Morocco itself (102,703 square miles compared with Morocco's 171,953 square miles). The armed forces of both neighboring countries are too small to permit any sustained offensive action in such a large area.[5] Even if they could spare the troops, their supply lines would be too long and too vulnerable to interdiction by air strikes.

A more likely scenario would be for one or both countries to give local guerrillas support and safe haven on their soil near border areas. A guerrilla campaign of harassment might keep the issue alive and prod Madrid into some overreaction that would tie down the Spanish contingent there. Spain and Morocco have already clashed over a related issue. On 2 April 1973 a Spanish fighter plane exchanged fire with a Moroccan coast guard frigate after the vessel

---

[3] Ibid., 22 August 1974.

[4] *Washington Post*, 24 May 1975.

[5] Morocco has a 50,000-man army, and Mauritania's total armed forces including paramilitary troops number 2,900.

intercepted a Spanish trawler fishing in disputed Atlantic waters. In March 1973, Morocco extended its fishing rights from 12 to 70 miles which included an area long regarded by Spain as its own fishing grounds for sardines and tuna.[6] More recently on 10 and 12 May 1975 ten Spanish soldiers including three officers were captured by guerrilla groups. Though Spain protested the action to Morocco, it is not clear that Morocco was responsible because Libya finances and arms a Saharan guerrilla group.[7] However, any significant increase in tension between Morocco and Spain over Spanish Sahara would probably be part of a much larger, more volatile dispute involving the Spanish enclaves on Morocco's northern coast.

Ceuta and Melilla. The two ports, Ceuta and Melilla, and the small island dependencies, Penon de Valez, Penon de Alhucemas, and Chafarinas, are fortified enclaves on the north coast of Morocco over which Spain has exercised full sovereign rights for several centuries. Morocco's claims to them rest on the same grounds as its claims to Spanish Sahara, that is, its rule of these areas in medieval times. Although Spain has been amenable to allowing self-determination in Spanish Sahara, it has been reluctant to consider any changes in the status of these ports and islands because of the overwhelmingly Spanish composition of their populations. Morocco has compared this situation to that of Gibraltar and has pointed to the inconsistency in Spanish policy. However, Morocco has placed much less priority on acquisition of these areas than on Spanish Sahara, though the issue is still active. At an April 1975 meeting of the Arab League Council composed of the foreign ministers of Arab countries, the organization renewed its support of Morocco's claim to these areas.[8]

Both Ceuta and Melilla are well-fortified garrisons where 8,000 and 9,000 army troops respectively are permanently stationed. These could be easily reinforced from the Spanish mainland. However, both ports are much more vulnerable than Spanish Sahara. Morocco could stage air strikes from nearby bases, artillery batteries could be set up within range of Ceuta and Melilla, and both could be harassed by ground troops. If just the two countries were involved, Spanish force could overwhelm Morocco. But such a conflict would not occur in a vacuum. A storm of Arab protest, a certain oil embargo against Spain, the likelihood that Algerian forces would aid Morocco, and

---

[6] *New York Times*, 4 April 1973.

[7] See the *Washington Post*, 24 May 1975, and the *Economist*, 24 May 1975.

[8] *Washington Post*, 29 April 1975.

the dilemma such developments would pose for the United States and the Soviet Union are all complicating factors.

While King Hassan is in power it is unlikely that Morocco will make such a bold move against the Spanish territories, but another Moroccan government might take a radically different attitude, particularly a regime headed by persons of the same orientation as the young officers who ousted King Idris of Libya in 1969. Given the history of political violence in the Mediterranean, such an abrupt change of government could occur at any time. The first of two recent attempts against the Moroccan monarchy took place on 10 July 1971 when rebel army troops broke into the king's summer palace and killed 93 guests attending his birthday party, though the king escaped unharmed. Because Libya's Radio Tripoli supported the rebels, Morocco broke diplomatic relations with Colonel Qaddafi's government on 14 July.

Thirteen months later on 16 August 1972, Moroccan air force jets attacked the king's plane while he was returning home from a trip to France and Spain. After his plane landed, the jets strafed the tarmac during a welcoming ceremony and attacked the royal compound after the king arrived at his palace. Hassan escaped injury in all three incidents which occurred within several hours. Two of the rebel pilots flew to Gibraltar and asked British authorities for asylum. This request was denied and the pilots were returned the next day to Morocco for trial and eventually were executed. On 19 August the British embassy in Tripoli, Libya, was attacked and severely damaged by demonstrators to protest the British action.

Gibraltar. The British decision concerning the rebel officers appears to have been based on the fact that without access to Morocco, Gibraltar would be virtually cut off. There is no traffic between Spain and Gibraltar, which makes Tangier a major transshipment point for essential supplies and personnel. Gibraltar has been a sensitive problem between Britain and Spain since it was captured by England in 1704 during the War of the Spanish Succession; it was formally made a colony in the Treaty of Utrecht of 1713. The Spanish position is that Gibraltar is an integral part of Spain and that the nature of present day strategic planning has diminished Britain's need for a base there. The British argue that the Gibraltarian community has a distinct identity and that, since a referendum in 1967 indicated an overwhelming rejection of any transfer of sovereignty to Spain, Gibraltar should remain under some sort of British aegis.

Despite the apparent rigidity of these positions and a move by Spain in 1969 to cut its links with Gibraltar, the protagonists in this conflict have engaged in limited negotiations. After a U.N. resolution on Gibraltar on 18 December 1968 which declared the "colonial situation in Gibraltar" to be "incompatible with the U.N. Charter" and called for the negotiated "termination" of this relationship, Spain initiated various efforts to bring about a negotiated transfer of sovereignty.[9] The British reaction to the United Nations resolution, as stated by Lord Caradon, was that "it will not and cannot be put into effect" and that this attempt to hand over the Gibraltarians against their will to Spain was "happily so far removed from possibility as to be incredible." [10]

In terms of relative power, Spain possesses superior force if it wants to use it. The British contingent in Gibraltar consists of just one infantry battalion, about 800 men, a squadron of outdated Hunter jet fighters, and a Royal Navy frigate. (Unlike other British forces which are being withdrawn from the Mediterranean, the Gibraltar garrison will be kept intact for the foreseeable future.) [11] Even though such a small force could easily be overwhelmed, Spain could not occupy Gibraltar without risking an armed confrontation. Both countries, however, maintain diplomatic relations and have sought to keep their differences over Gibraltar at a low level while maintaining their basic positions, including the Spanish land blockade. The outlook is complicated by one unknown factor: the future course of Spanish politics.

Spain after Franco. By mid-1975, the era of Spain under Generalissimo Francisco Franco, who was then 82, had not ended, though reports suggesting that he might retire soon were beginning to crop up in the rigidly controlled Spanish press.[12] It is assumed that when Franco does leave power his intended successor as head of state, 37-year-old Prince Juan Carlos, will gradually liberalize the Spanish political process—although the release of the many forces that have been held in check by Franco's presence may make a smooth transition impossible. At the very least there will probably be some crisis of succession as Juan Carlos tries to reestablish the monarchy and get a firm grip on the government while dealing with such divergent groups as the Roman Catholic Church, the army,

---

[9] *Times* (London), 19 December 1968.

[10] Ibid.

[11] *Statement on the Defence Estimates 1975* (London: Her Majesty's Stationery Office, 1974), p. 14.

[12] *Washington Post*, 7 May 1975.

Basque separatists, labor, the National Movement (Spain's only legal political organization), the now outlawed Communist party, and Spain's numerous anarchists. Even under Franco it is impossible to completely control these factions. For example, the Basque separatist movement (ETA) claimed responsibility for the assassination on 20 December 1973 of Prime Minister Admiral Luis Carrero Blanco.

At worst, a violent release of pent up emotions could result in another crippling civil war, making Spain a liability to the West. The lesions on Spain's body politic are still raw from the conflict of the 1930s and Spanish industry and agriculture are just recovering from its devastating effect. The appearance of an ultra-conservative government after a transition period could be damaging to American interests. Such a government might want a close military relationship with the United States but would probably seek higher terms in the areas of arms assistance and sales, mutual security arrangements, and diplomatic support in exchange for allowing a continued American military presence in Spain. A significantly expanded commitment or a sizeable increase in military sales to an ultra-right-wing Spain would probably be unacceptable to the American Congress.

A leftist regime in Spain on the model of its neighbor, Portugal, would be equally bad for the United States. In this scenario, all American bases in Spain would be closed, Spanish Sahara and the enclaves would be given independence or transferred to Morocco, and Britain would be given an ultimatum to leave Gibraltar. A pro-Communist regime in Madrid might also try to reach an agreement with Morocco to regulate shipping through the Strait of Gibraltar, restricting the passage of warships of non-Mediterranean nations or preventing such vessels from carrying nuclear weapons. The United States might choose to ignore any limitation on its right of free and unannounced passage through the strait, but its position in the Mediterranean would be hurt by the necessity of forcing its way into the basin.

At first glance so extreme a move to the left may appear a bit farfetched. But there have already been a variety of proposals to restrict the navies of the two superpowers and to make the Mediterranean a nuclear-free zone.[13] It must be remembered that only two

_____
[13] Several countries have called for a policy of "the Mediterranean for Mediterraneans," but only a few have adopted a hard-line position. Tunisia, which for two years had been advocating the withdrawal of all foreign navies from the Mediterranean, reversed itself in April 1973 when the U.S. Sixth Fleet gave a great deal of assistance to that flood devastated country (*New York Times*, 8 April 1973). Other countries are, in some sense, resigned to the fact that "they have to live with the superpowers' navies within aircraft and missile range for a long

years ago few analysts would have thought it likely that Portugal would soon be ruled by a leftist group of army officers who would unilaterally free the Portuguese colonies in Africa.

The Future of Malta. The planned withdrawal of British forces from Malta by 1979 may be a crisis on the horizon. If the West does not carefully and thoughtfully formulate a policy for Malta through the end of this decade and through the 1980s a confrontation similar to that of 1971–72 could occur. But the next time around Malta may not favor the West.

The last crisis over the future of the British bases in Malta provides a useful study in miniature of how a basically economic problem can quickly become a larger dispute involving diplomatic and strategic interests. It is also a poignant illustration of the myopic character of Western policy in the Mediterranean.

The Maltese archipelago, consisting mainly of the island of Malta, 17 miles long and 8 miles wide, and the smaller island of Gozo, 9 miles by 5 miles, has never been self-sufficient. This is not to say that Malta did not prosper; it did as a colony in the British Empire. Under the British, Malta was a fortress, a strategic link to the empire east of Suez and a symbol of Western naval power in the Mediterranean. But this prosperity depended on one factor—the British. The British government made direct payments for base rights; British servicemen spent their money in the bars and on the girls of "The Gut"; large numbers of Maltese were employed at British installations; goods and services sold to the families of British servicemen generated additional employment and income; many British retired there; and British tourists found the Mediterranean sunshine a welcome relief from the dank weather of England.

---

time" (*New York Times*, 11 November 1973). Even among the NATO allies, however, this resignation has not always meant cooperation.

The only formal proposal appears to have been a resolution introduced at the Fourth Conference of Non-Aligned Nations held in Algiers in September 1973. Without specifically mentioning any fleet, the resolution called for the withdrawal of all foreign navies from the Mediterranean (*New York Times*, 9 September 1973).

Beyond this, there has been much discussion of making the Mediterranean a nuclear-free zone. Spain and Yugoslavia have voiced concern about the presence of nuclear weapons (*New York Times*, 11 November 1973). The Soviet Union has also called for the removal of nuclear arms from the area. Secretary Brezhnev, in speeches delivered in Poland, July 1974, and East Germany, October 1974, suggested that both the United States and the U.S.S.R. withdraw all nuclear armed ships (including submarines) from the Mediterranean (*New York Times*, 7 October 1974).

When Malta obtained full independence in 1964, the British Empire had already been replaced by the Commonwealth and British power east of Suez and in the Mediterranean was on the decline. As the British presence diminished, the hole in Malta's economy became larger and more obvious. About 70 percent of Malta's food must be imported. Malta has no known mineral resources other than limestone and salt, and it has no hydroelectric potential. When the British presence shrank, unemployment rose to 8 percent. (There is a work force of about 100,000 out of a population of 319,000.) Two five-year plans to convert the former British naval dockyard to commercial ship repair facilities were only partially successful because the 1967 Arab-Israeli war left the Suez Canal closed. Since ships no longer used the Mediterranean route to the Persian Gulf, East Africa, and beyond, the volume of shipping that Malta was hoping to attract to its repair facilities was greatly reduced. Several Maltese politicians, notably Dom Mintoff, were alarmed by this. They felt that Britain and NATO ought to pay more for their facilities on the island. They also wanted to modify the economy to make it more diversified and, if possible, self-sufficient. Mintoff, who was prime minister from 1955 to 1958, had expressed this view many times in public speeches and press interviews. So when his Labor party won a one-seat majority in June 1971, he made getting more money from the West and improving the economy his first order of business. On 30 June, he abrogated the Anglo-Maltese Defense Agreement of 1964, an action which the Western press treated with surprise and shock as it was encouraged to do by Western governments.

The West, especially Britain, the United States, and Italy, was aware that Mintoff had a good chance of becoming prime minister again; they knew his mood and they understood Malta's real economic problems. They were also conscious of Malta's "negative strategic value" to the Western alliance. This convoluted phrase was invented during high-level NATO policy discussions in 1964 and summed up several developments. Malta's importance was no longer what it had been in Nelson's day or during World War II. If the NATO facilities based in Malta were moved, it would not make a great tactical difference to the alliance. The British Nimrod and Victor maritime surveillance aircraft could do their jobs equally well based in Sicily. NATO radars could track shipping from Sicily and Pentelleria. But naval bases on Malta would have enormous strategic value to the Soviet Union—and this would have a "negative" impact on NATO.

Had Britain or NATO made a proposal to explore ways of improving Malta's economy and discussed an increase in rent—which eventually was implemented anyway—a crisis probably would have been avoided. But nothing was done and the initiative passed by default to Mintoff, who used it ruthlessly. He threatened to end summarily the British presence unless more money was paid, he hinted that "foreign" armed forces would be brought to Malta without specifying their nationality, he made several trips to Libya for talks with Qaddafi giving the impression that Libyan money was available if the British payments ended, and he received with wide publicity the Soviet ambassador to London, who is also accredited to Malta. All of this put NATO embarrassingly on the defensive at a time of growing Soviet naval power and influence in the Mediterranean.[14] On 20 August 1971 at Mintoff's insistence, NATO ordered the transfer of its Mediterranean naval headquarters from Malta to Naples. Though this was just an administrative shift, it was significant. "The implication was clear," the International Institute of Strategic Studies said in commenting on the crisis, that "if NATO did not want the facilities badly enough to pay Mr. Mintoff's price, it might find the Soviet Union there instead." [15]

The crisis need not have occurred, but once the dispute had arisen, it should have been solved without the harsh atmospherics, though Mintoff's behavior was trying at times. It is tempting to say that the British were largely to blame. They, of all the NATO countries, certainly were the best placed to see the situation brewing and respond to it. But ultimately this was a collective Western failure. While the dispute was heating up, the United States, out of respect for its closest ally, followed the British lead in dealing with Malta—which meant taking no initiatives that might offend Britain. The American deference to Britain was so complete that President Nixon and British Prime Minister Heath did not discuss Malta during their meeting in Bermuda on 20–21 December 1971.[16]

It wasn't until after Mintoff had announced that 15 January 1972 would be the final deadline for the withdrawal of British troops, and Britain had said it had no intention of meeting it, that the United States stepped in. President Nixon made a trans-Atlantic

[14] At this time Western navies in the Mediterranean also were dealing with the reconnaissance flights of TU-16 Badgers, jets with Egyptian markings but flown by Soviet pilots from Egyptian airfields.

[15] *Strategic Survey 1971* (London: International Institute of Strategic Studies, 1972), p. 30.

[16] Malta, however, was discussed during the parallel talks between Secretary of State Rogers and Foreign Secretary Douglas-Home, who were also in Bermuda.

telephone call to Heath to explore ways of solving the crisis, including a NATO contribution to an increased rental payment. In the meantime Mintoff had asked the American embassy in Sliema for the private telephone number of the White House. He did not get through to Nixon or his national security adviser, Henry Kissinger, but he did talk to Robert Ellsworth, then the U.S. ambassador to NATO. The British were annoyed even at this low-level response, feeling that it served to undercut their negotiating position with Mintoff.

It is difficult to conceive how the situation could have deteriorated further. Rumors of impending violence swept the island. Noisy demonstrations in Valletta contributed to the crisis mood. Britain began evacuating service families as British pensioners abandoned the island. A group of 44 Libyans and Egyptians arrived on the island. Most were there to operate the control tower of the Royal Air Force base which also served as Malta's only civilian airport. There were at least two intelligence officers in the group, which caused alarm at the British High Commission and U.S. embassy. They also brought a supply of ammunition, apparently for use by the Maltese Land Forces, Malta's 688-man army. (It turned out that the ammunition would not fit the MLF weapons.) London papers were filled with pictures of British women clutching their children as they left the island and stories about lifelong pets having to be "put down" (an American would say "put to sleep") because health regulations prevented them from entering Britain.

After intensive consultation with the NATO allies, serious talks between Mintoff and British Defense Minister Lord Carrington began before the showdown date of 15 January. On 26 March Malta and Britain signed a new seven-year defense agreement. The accord allows the stationing of British forces in Malta for Britain's and NATO's defense and specifically excludes Warsaw Pact forces from using the island's facilities. Britain and NATO agreed to pay Malta $36.4 million in annual rent, with the United States paying $9.5 million of this. Another $23 million in bilateral aid from NATO countries was also promised to Malta. "Negative strategic value" had won the day.

Malta's economic future now is somewhat brighter. The Suez Canal is open again and Malta's extensive ship repair facilities and deep natural harbors are ideally located to serve the revived Mediterranean shipping. China, Libya, and several Arab countries in the Persian Gulf have given financial help to modernize the shipyards and begin construction of supertankers.[17] Chinese and Libyan assistance notwithstanding, the West ought to continue to play a substantial role in Malta's economy, if for no other reason than to provide a viable alternative to Soviet or other anti-Western influence.

Zone B. The dispute between Italy and Yugoslavia over Zone B is a complex and potentially volatile issue involving a 200-square-mile area on the Istrian Peninsula south of Trieste which came under Yugoslav control after World War II. It contains no large population centers, the principal towns being Kopar (Capodistra), Piran (Pirano), and Novigrad (Cittanova). The population of Zone B is predominantly Slovene, with the Italian minority concentrated in the towns. Zone A, which contains the city of Trieste, is predominantly Italian and is currently under Italian sovereignty. Both zones were formerly part of the Austro-Hungarian Empire. Following the completion of rail links to Vienna and the Danube Basin in 1857 and the subsequent opening of the Suez Canal, Trieste became the major port of Central Europe.

The territory of Trieste and the Istrian hinterland was promised to Italy in the Treaty of London in return for Italy's entering the war against the Axis powers. Italy acquired formal possession of these areas by the Treaty of Rapallo in 1920, despite the objection of Woodrow Wilson that the large Slovene population of the area deserved the right to national self-determination. During the interwar period, with the rise of fascism in Italy, the Slovene minority was ruthlessly repressed. At the same time, the importance of Trieste as a port declined as Yugoslavia rerouted its trade to its own port cities. At the close of World War II, Allied troops moved into Trieste from the west while Yugoslav troops entered from the east, filling the vacuum left by the retreating German army. Yugoslavia justified its attempt to gain control of the entire region as an effort to "liberate" Trieste, which it felt should come under Yugoslavian sovereignty. The Allies resisted this effort, with Britain and the United States insisting that the city of Trieste itself, because of its predominantly Italian character, should remain under Italian control.

After much bitter debate, the opposing parties finally agreed to a compromise which called for establishing the territory of Trieste and the Istrian Peninsula as an independent state to be named the Free Territory of Trieste. Under this agreement, military occupation would continue temporarily, with the area occupied by British and American troops designated Zone B. Military rule over the Free Territory of

---

[17] *Washington Post*, 2 January 1975.

Trieste would end when a governor was installed, at which time responsibility for the integrity of the new state would rest with the United Nations Security Council.

For a number of reasons, primarily the failure of the Security Council to agree on a suitable governor, this plan was never put into operation, and Zones A and B reverted to Italy and Yugoslavia respectively, as specified by the London Memorandum of 1954.[18] This memorandum was a de facto settlement only beacuse the Security Council resolution setting up the state of Trieste was never formally deactivated. Though Italian and Yugoslav officials claimed that they had made great sacrifices in reaching this accord, both sides welcomed the settlement. Subsequent problems derive from the fact that, while Yugoslavia takes the position that this arrangement is permanent and that Zone B no longer exists, Italy has not formally relinquished its claim to the entire territory.

Within Italy, Communist and neo-fascist elements objected to the London Memorandum compromise from the beginning. The neo-fascists figured in one recent flare-up of the dispute, when the Zone B question led to a two-year postponement of an official visit to Italy by Yugoslav President Josip Tito originally scheduled for December 1969. Charges by representatives of the neo-fascist Italian Socialist Movement (MSI) in the Italian parliament that formal concession of Zone B was to be a topic of discussion during the visit forced the Italian prime minister to respond publicly that "there could be no concessions in matters of the national interest."

A more serious flare-up developed in early 1974 when Italy presented a note verbal to the Yugoslav ambassador to Rome which referred to Zone B as Italian territory.[19] Under pressure from Slovinia, and anxious to capitalize on the domestically unifying effect of such incidents, the Yugoslav government responded to the Italian statement with a virulent propaganda campaign against what appeared to be the "irredentist and fascist aspirations" of Italy.[20] The Yugoslav reaction culminated in the mobilization of the army, after which the government allowed the situation to cool off, having gained sufficient propaganda benefit. One further explanation of the intensity of the

---

[8] "Memorandum of Understanding between the Governments of Italy, the United Kingdom, the United States and Yugoslavia regarding the free territory of Trieste," London, 5 October 1954.

[9] English-language dispatch of the Yugoslav press agency, Tanjug, 4 April 1974, quoting the text of the Italian note verbal of 15 February 1974, delivered 21 February 1974.

[0] Text of verbal reply of Yugoslavian ambassador to Italy's note verbal, which was widely broadcast and distributed by Tanjug on 4 April 1974.

Yugoslav reaction lies in the emotional importance of the Zone B and Trieste issue for the aging Tito, who would have gone to war over it in 1945–48 if he had had Soviet support, and who hopes to settle the dispute once and for all before his death. Addressing the Tenth Party Congress of the Central Committee of the League of Communists of Yugoslavia in May of 1974, Tito spoke about the threat this dispute poses to Yugoslav-Italian relations:

> In contradiction of the spirit and practice of neighborly rela-
> tions and long years of cooperation during which enviable
> results were achieved, the Italian Government recently
> voiced open territorial pretensions towards Yugoslavia. This
> is a direct attack on the sovereignty and territorial integrity
> of this country. We consider the border question non-
> existent. We can no longer negotiate on this matter.[21]

Italy's chronic malaise. It is unlikely that the Zone B dispute will result in an armed confrontation as long as Italy is governed by coalitions of relatively moderate political parties. But there is no assurance that this situation will continue; public dissatisfaction with the inability of these traditional parties to deal with the country's pressing and recurring political, social, and economic problems is growing. Since the end of World War II there have been 37 govern- ments in Italy; crippling strikes by both government and nongovern- ment workers frequently bring major cities and essential services to a standstill; the inflation rate for 1974 was 25 percent; terrorism by extreme left- and right-wing groups plagues the country; and Italy's Communist party is the largest in Western Europe and the second largest political group in the country.

Against this background of domestic instability Italy faces un- certainties in the international arena. It is dependent on Arab oil for its energy; Britain's plans to withdraw its forces from Malta by 1979 will leave a potentially dangerous vacuum on Italy's doorstep; the Suez Canal's reopening is both an opportunity and a cause for concern; and Portugal's abrupt slide to the left raises doubts about its continued participation in NATO. Probably the international devel- opment that most worries Italy, however, is what will happen in Yugoslavia after Tito. The pressures within Italy generated by the pres- ence of a pro-Soviet regime in Yugoslavia just across the Adriatic Sea

[21] "Changes in the World Today and Yugoslavia's Foreign Policy," Report the Central Committee of the League of Communists of Yugoslavia (CCLC to the Tenth Party Congress, delivered by Josip Broz Tito, General Secreta of the CCLCY, on 27 May 1964, reprinted in Jeffrey Simon, *Ruling Commun Parties and Détente: A Documentary History* (Washington, D. C.: Americ Enterprise Institute, 1975), p. 226.

102

would probably be more than Italy's fragile system could comfortably accommodate. The Communist party, whose prestige would be greatly enhanced by this development, probably would become part— more likely a leading member—of the government. Or, conversely, there could be a slide toward the extreme right, perhaps even a coup d'etat, as a reaction against a pro-Soviet Yugoslavia.

Post-Tito Yugoslavia. Concern over what happens in Yugoslavia after Tito is not confined to Italy. Like Franco's in Spain, Tito's passing from power is another inevitable change of leadership that could occur at any time. It could produce a serious international crisis and a major shift in the Mediterranean balance of power in addition to the purely domestic political transition process. On 25 May 1975, Tito was 83 and though he is in surprisingly good health for a person of his age, his physical condition is deteriorating.

Implicit in the administrative structure that Tito has set up— a collective leadership based on rotating chairmen of a federal council—is that no single leader will be strong enough to follow him. Ethnic nationalism has always been strong in Yugoslavia, most notably among the Serbs and the Croatians. Some Croatians, whose national capital is Zagreb, have exhibited strong pro-Soviet sentiments in the past. One exile publication, *Hrvatska Drzava* (The Croat State), once suggested offering to give the Soviet Union air and naval bases along the Adriatic if Moscow would underwrite an independent Croatian state.[22] Croatia borders Hungary where the Soviet Union has four divisions stationed. Tito has been the cement keeping the ethnic blocs in place. In foreign affairs Tito follows a nonaligned policy, keeping a foot in both the Western and Eastern camps. He has accepted economic and military aid from the United States, offered as an incentive to stay out of Moscow's satellite system. Yugoslavia is a member of the Council for Mutual Economic Assistance (COMECON), the Moscow-sponsored trade and economic organization, but also maintains ties with Western Europe's Common Market.

It is an open question whether the rotating federal council can maintain—or would want to maintain—this policy of balancing between East and West. Yugoslavia's break with Stalin in 1948 was largely a matter of Tito's personality buttressed by the fierce personal loyalty he had commanded during the World War II resistance movement. Yugoslavia was the only country in Eastern Europe that

Yorick Blumenfeld, *Yugoslavia in Flux* (Washington, D. C.: Editorial Research Reports, Congressional Quarterly, 1973), vol. 1, p. 425.

liberated itself from German occupation rather than waiting for th arrival of the Red Army. No comparable leader is on the horizon Yugoslavia today.

Even though Tito has set up machinery that he hopes will ensu a smooth transition, he cannot ensure that the Soviet Union will n try to foment an ethnic crisis and then attempt to move troops and end the chaos in the name of restoring order on its Balkan flan It is also conceivable that both American and Soviet forces could g involved in a confrontation over Yugoslavia after Tito leaves t scene. Yugoslavia more than Portugal or Spain would be a test détente. It is one thing for the Soviet Union to support a Commun party in Western Europe, though the United States ought to oppo and try to frustrate such an effort. It is quite another matter for t Soviet Union to covertly use political and paramilitary action to al the strategic balance in the Mediterranean.

Moving eastward we encounter the dispute over Cyprus and t Arab-Israeli conflict. We have already examined how these situatic affect the policies of the two superpowers and the deployment of th forces. In the context of this study there is little that can be said tl is not highly perishable because these situations are so dynamic. addition to these problems there are other outstanding disputes the eastern Mediterranean Basin that deserves discussion—the quar between Turkey and Syria over Alexandretta and Lebanon.

Alexandretta. With its mixed Turkish-Arab-Kurdish population, *sanjak* of Alexandretta,[23] which Turkey calls Iskanderun, has b a bone of contention between Turkey and Syria since the conclus of the Franco-Syrian Treaty of September 1936 which promised Sy independence. Under that treaty Alexandretta was considered to be integral part of Syria. Turkey, which had considerable interest in *sanjak* on account of its predominantly (40 percent) Turkish popt tion, protested. The basis for these protests was the sudden cha in the status of Alexandretta which had formerly enjoyed se autonomy under the terms of the 1921 Franklin-Bouillon Treaty.

Turkey brought the issue before the League of Nations Cour which drew up a special statute renewing the autonomy of the *sar* and assigning various responsibilities to France, Syria, and Turl Alexandretta was thus virtually separated from Syria. Despite new arrangements and the Syrians' efforts to redress their posit Turkish agitation regarding the *sanjak* continued, culminating in Turkish denunciation of the Turko-Syrian Treaty of Friendshi

---

23 *Sanjak* is an Ottoman Turkish administrative term.

1926 and an anti-French campaign. France, in search of allies in the Mediterranean, found the Turks to be more valuable and reliable than the Syrians and was prepared to accommodate Turkey on this issue. On 3 July 1938 France and Turkey reached an agreement that established the *sanjak* as a Franco-Turkish condominium until general elections could determine its final status. In the presence of Turkish troops the election produced a 22-out-of-40-seat majority for the *sanjak* Turks, who proceeded to proclaim autonomy under the new name of the Republic of Hatay. The final severing of Alexandretta from Syria was thus duly notarized.

Although this issue has remained essentially dormant since 1939, the Syrians still consider Alexandretta part of Syria. Their interest in the *sanjak* survives because of its sizable Arab population, the refugees who fled Turkish rule, and its geographical confluence with Syria proper. Indeed, a glance at the map may suggest a certain validity to the Syrian claim. Because of the current primacy of the Arab-Israeli conflict in the Syrian national consciousness, this issue is now in the background. It has not, however, been forgotten.

Lebanon's Future. Since the 1967 Arab-Israeli war, two important and related developments have occurred in Lebanon. One is the growth of the political power of the more than 350,000 Palestinian refugees living there. The second is the open restlessness among Lebanese Moslems about their traditional political inferiority.[24] The 1973 war accelerated both developments. The results are that Lebanon's delicate balance between Christians and Moslems has been upset, the orientation of its economy toward banking and other services is in jeopardy, and its status as a bridge between the West and the Arab East has been seriously eroded. Fighting in Beirut during April and May 1975 between right-wing Christians and Palestinian guerrillas polarized the two religious communities and shattered the calm that has been the main international attraction of Lebanon, an oasis in the midst of Middle East violence and uncertainty.

In addition, Lebanon may get caught in the middle by pressures from Syria to reform its confessional system to give Moslems a more equitable share of political power and pressures from Israel to exert control over Palestinian guerrillas operating from bases in southern Lebanon. The West no longer has to rely on Lebanon's good offices to conduct political and economic business with the Arab world. Western firms now have their Middle East offices in the countries or

---

4 *New York Times*, 12 March 1975.

regions, like the Persian Gulf, where they do business. Negotiations to settle the Arab-Israeli conflict have moved to a new stage. Egypt and Syria speak for themselves, if not directly to Israel, through a much better peace broker, the United States. In short, Lebanon is in danger of being overtaken by events in the Middle East. Its potential enemies, in any realistic scenario, are Syria and Israel. It is even conceivable that there could be a simultaneous invasion of Lebanon by both countries: Israel pushing up through south Lebanon to the Litani River just above Tyre, and Syria sending troops down the coast to Tripoli and across the anti-Lebanon range into the Beka'a Valley. For Lebanon to create an army, air force, and navy to guard against such a threat would require a radical change of its national policy. It probably could not be done. It would simply cost too much in political, social, and economic terms.

Lebanon's 14,000-man army is more suited to helping maintain the domestic peace. It is used, for example, to police the polls during national elections. But because the army is so small, balloting is held over several consecutive Sundays in different parts of the country to give the army time to move to the different constituencies where the voting takes place. The army must also face the constant security problem posed by the Palestinian refugee camps, which are spread throughout the country and which are guerrilla strongholds. Power ful clans built around family groupings that maintain heavily armed feudal armies equipped with such weapons as the AK-47 Soviet-made automatic assault rifle also defy government control.[25]

To try to combat the external threats, Lebanon has historically turned to the big powers and the United Nations. The U.S. marine landing in July 1958 stopped Syria's heavy involvement in the civil war. But American pressure and world opinion have not prevented Israeli raids into southern Lebanon against Palestinian guerrillas who use the area as a base for attacks on Israel. These operations—both the Palestinian attacks and the Israeli retaliation—are an acute em barrassment to the government in Beirut and constantly threaten to undermine its authority. For domestic and intra-Arab political rea sons the army does not attempt to clear the guerrillas from their strongholds in the south, and for military reasons Lebanon's armed

---

[25] For example, in late February 1975 the Beirut government awarded a fishing contract to a large commercial firm. Local fishermen in the southern town of Sidon who use small rowboats and nets that they can handle on their own felt threatened by the decision. The violence began with a demonstration by the fishermen that the army tried to quell. It escalated into a fight involving auto matic weapons and rocket fire. At least 20 persons were killed in the clashes. *Washington Post*, 27 February 1975.

forces cannot deter—or even engage with any hope of success—the invading Israeli forces.

Serious and widespread domestic violence that would pit Moslem against Christian and involve the Palestinians would probably follow a move against the guerrillas. Lebanon would also antagonize Syria and might provoke a military reaction if it moved against the guerrillas on a large scale. Even without the threat from Syria, it is difficult to believe that Lebanon's army could do what the Israeli army has not accomplished in southern Lebanon during the past five years. Lebanon's tiny army and 14-plane air force would be overwhelmed in such a situation. In the face of Lebanon's inability to check both a threat from Syria and violence near its frontier, Israel could decide to move into southern Lebanon itself. (Hence the scenario for simultaneous invasion.)

The necessity of at least appearing to deal with the Israeli raids and to establish control in the south has caused a few changes in the army and has led to a search for more modern weapons. The army has deployed more men to the south, but rarely engages Israeli forces, primarily because it does not have the equipment to fight on the spot or furnish back-up support if Israel escalates the conflict. Recently Lebanon acquired TOW wire-guided antitank missiles from the United States. Lebanese officials have also discussed getting Vulcan 22mm. antiaircraft guns, M-48 tanks, and A-4 Skyhawk fighter-bombers from the United States. These arms talks are a direct result of the Israeli raids into southern Lebanon.[26] There is also a possibility that Lebanon may acquire some Soviet-made antiaircraft missiles from Syria or buy similar weapons from France.[27]

The big question is whether these new weapons will be used. If Israel thought they were a real threat to its fighter-bombers, the missile sites could easily be rendered ineffective before strike aircraft were sent over. A more suitable aircraft for Lebanon might be helicopters—a mixture of gun ships, such as the AH-1G Huey Cobra, and armed troop carriers. Lebanon is only 125 miles long and 50 miles wide at the deepest point. A supersonic jet would overfly the frontiers in a matter of minutes at full speed. Helicopters are able to direct firepower with more precision than fixed-wing jet aircraft.

Rather than purchase token quantities of modern weapons, a more urgent task for Lebanon might be to restructure its political system. The unwritten National Pact of 1943, on which the country's confessional—or religious proportional—system is based, calls for a

---

26 Ibid., 17 January 1975.
27 *New York Times*, 22 January 1975.

six-to-five Christian-Moslem ratio in public offices, the unicameral legislature, the military, and the civil service. The ratio is based on Lebanon's last official census in 1936 and strengthened by a more recent citizenship law that requires Lebanese immigrants, most of whom are Christians, and their descendants to be counted in any future census.

Under this arrangement, the president, the foreign minister, and the commander of the armed forces are Christians. The premier and speaker of the legislature are Moslems. Membership in the parliament is always a multiple of 11 and there is always a six-to-five Christian majority. This means that the foreign and defense policies are guided by Christians and that Moslems form a perpetual opposition. There is general agreement that there are now more Moslems than Christians among the estimated population of 3.2 million. However, any suggestion that a political adjustment ought to be made to account for population shifts is viewed as threatening by right-wing Christian politicians. The task will be all the more urgent if there is a settlement of the Arab-Israeli dispute and Lebanon moves even further away from center stage vis-à-vis the West to become the focus of the scrutiny of its Arab neighbors, particularly Syria.

An Independent Palestine? The prospect of a Middle East settlement raises the possibility that a separate Palestinian state on the West Bank of the Jordan River and in the Gaza Strip on the Mediterranean might be part of the peace package. The Palestinianism movement is already a new element in Mediterranean politics. If a Palestinian state is created it will add another dimension to this development. In its infancy—probably through the end of this century—its most pressing business will be organizing itself and settling the Palestinian diaspora.

The predominant foreign influence will probably be Arab, with Cairo, Damascus, Amman, and Tripoli competing for power while Israel watches anxiously. In the larger arena, the Soviet Union because of its arms and diplomatic support of the movement will probably be viewed more favorably by a Palestinian government than will the United States.

The Suez Canal. On 5 June 1975, after being closed to shipping for eight years, the Suez Canal reopened with almost as much fanfare as accompanied its inauguration in 1869. As if to emphasize the strategic importance of the 103-mile waterway, President Sadat wore the uniform of Egypt's admiral of the fleet and sailed aboard an Egyptian destroyer, the lead ship in the first convoy.[28]

In strictly naval terms, the canal will permit warships of both the United States and the Soviet Union—and other powers—quicker access to the Indian Ocean and Persian Gulf from the Mediterranean. The distance from Norfolk to the entrance of the Persian Gulf using the Suez Canal route is 9,750 miles. Ships using the Cape route around Africa have a 13,750-mile trip to reach the gulf from Norfolk. But U.S. ships based in Subic Bay in the Philippines in the South Pacific have a much shorter distance to travel. For Soviet ships based in the Black Sea and the Mediterranean, the distance is 2,200 miles from the Black Sea to the Indian Ocean using the canal. Ships making the same trip going around Africa would have to travel 11,400 miles. Soviet ships based at the Pacific port of Vladivostok have a 9,000-mile voyage to the Indian Ocean.

These differences in distance are frequently cited by the Pentagon to indicate that the Soviet Union gains the greater advantage from having the canal open.[29] Such comparisons overlook other considerations. First, all ships stationed in the Mediterranean have the same advantage of access to the canal. Before the canal closed in 1967, the Sixth Fleet supplied the two destroyers that rotate to the Middle East Force, the U.S. Navy unit homeported in the Persian Gulf island of Bahrain. It was only after the canal closed that the ships had to make the long trip around Africa from Norfolk. The second consideration is that the Soviet Union is restricted from using the canal as a reinforcement route by the size and crisis role of its Black Sea fleet and the Fifth Escadra. It would not want to risk depleting either by taking ships from these units on other missions, rushing them through the canal to the Indian Ocean. It would only be a marginal advantage if the Soviet Union shifted naval forces to the Mediterranean and Black Sea from other areas. Such a development would be immediately apparent to Western intelligence and would have ominous political significance. The Soviet Union could, over a period of two or three months, significantly increase its naval presence in the Indian Ocean—by about 20 ships—but it does not need the canal to do that.

---

28 Ibid., 5 June 1975.

29 See interview with Secretary of the Navy J. William Middendorf III in the *Washington Post*, 5 June 1975, the day the canal opened. Middendorf said, for example, that the Soviet navy will be able "to bring their whole fleet through. We can't bring our carriers through—a lot of our big stuff we can't bring through and we wouldn't wish to anyway." Also see "Opening of Canal to Alter Balance," *New York Times*, 19 November 1974.

In my judgment there is a greater strategic advantage for the United States. Egypt plans to dredge the canal to its pre-1967 depth of 38 feet which would permit—as it did before the canal closed— American warships including some smaller aircraft carriers to transit the waterway. During the ceremonies on 5 June, for example, the Sixth Fleet flagship, the U.S.S. *Little Rock*, was the only foreign warship to join the opening procession.

Of equal—perhaps greater—importance is what the reopened canal will mean in economic terms and as a stabilizing element in the Middle East. In 1966, its last full year of operation, the canal handled 21,250 ships including 10,000 tankers carrying 242 million tons of cargo, or 14 percent of all world sea trade for that year. It is estimated that the closing of the canal cost the West, Japan, and the countries of East Africa and Southeast Asia about $5.7 billion annually due to loss of trade, increased shipping expenses, and other economic costs.[30] Though the world adjusted to the closing by rerouting shipping traffic, altering trade patterns, and developing the supertanker, the canal may regain some importance to world shipping.

Egypt has invested a lot of energy in revitalizing the canal zone—rebuilding the cities of Port Said, Ismailia, and Suez, planning to enlarge the waterway so it can accommodate ships up to 200,000 tons, and preparing to dig a tunnel under the canal. These are indicators that it does not want another war with Israel and that it is willing to use the canal as a hostage to peace. That is the major significance of the reopened Suez Canal and one that, in my judgment, outweighs the ability of Soviet and American warships to reach the Indian Ocean more quickly.

Tunisia. There are fewer uncertainties about what will happen in Tunisia after President Habib Bourguiba leaves power than there are about what will happen in Spain and Yugoslavia. Bourguiba, who became 73 on 3 August 1975 and who is in poor health, is president for life, which means he probably will die in office. Under Tunisia's constitution he will be succeeded by the premier, Hedi Nouira. The ruling political party, Neo-Destour, is a stable organization that commands broad support in the country and from the 16,000-man military. The government that follows Bourguiba is expected to retain its ties to the West, particularly France and the United States, while continuing to play a role in the Maghreb and the Arab world.

---

[30] U.S. Department of State, Bureau of Public Affairs, "Suez Canal," *Foreign Policy Outline*, May 1974.

If a smooth succession is jeopardized, the source of the trouble is likely to be external, probably Libya. In January 1974 Tunisia accepted and then abruptly renounced a political union with Libya, causing great public embarrassment for Colonel Muammar Qaddafi whose plan for a union with Egypt met a similar fate. Libya reportedly has given covert support to several Tunisian dissident groups that may try to exploit the period of transition in the immediate post-Bourguiba era.

Other Power Elements

France. After the United States and the Soviet Union, France is the most influential nation in the Mediterranean Basin. And of all Mediterranean countries, France is probably the only one that has the requisite ingredients—the political and economic motivation, basic pro-Western orientation, internal stability, and military strength—to play a positive big-power role in the affairs of the region. Though all but a few of its overseas territories have gained independence, France maintains close ties with its former colonies and still sees a function for itself on the world scene. The Mediterranean Basin, which is a relatively small area and one where France has considerable experience as a big power, provides a convenient and manageable outlet for its foreign policy aspirations. France is the architect and principal advocate of the European Common Market's so-called "Mediterranean Policy" which is the effort by the EEC to knit a series of bilateral association agreements between itself and Mediterranean countries. France depends on foreign sources, mainly in North Africa and the Middle East, for oil and natural gas. It also wants to recycle the petrodollars it spends on oil and gas. This was the primary reason for the sale to Libya, beginning in 1969, of 110 Mirage fighters for an estimated $144 million. In addition to Libya, ten other Mediterranean countries use French military equipment.[31] France conducts a substantial amount of nonpetroleum and nonmilitary trade in the Mediterranean as well.

In the broad area of foreign and defense policy France is taken seriously and has credibility both because it is a nuclear power—the only acknowledged one in the Mediterranean—and because it has sufficient domestic cohesiveness to see its policies through. (Italy might also be in this category but for its chronic instability.) France

---

[31] U.S. Arms Control and Disarmament Agency, *World Military Expenditures and Arms Trade 1963-1973* (Washington, D. C., 1975), Table 3, p. 67.

possesses nuclear weapons and has the capability of delivering them by land-, air-, and sea-launched systems. It has applied nuclear technology to its navy: it plans a total of five nuclear-powered ballistic missile submarines, three of which were in service in 1975, and funds have been earmarked for two nuclear-powered attack submarines, and the construction of a nuclear-powered aircraft carrier to supplement the two conventionally powered carriers now in service was to begin during 1975. It is also the only Mediterranean country whose navy includes fixed-wing aircraft carriers, which were scheduled to be transferred from the Atlantic port of Brest to the naval base at Toulon in the Mediterranean in 1975.[32]

The question is whether France will conduct its foreign policy in concert with American interests in the Mediterranean—or at least pursue objectives that are not inimical to American interests. At times France has been a trying partner to the United States in the Western alliance. It remains a member of NATO, but does not participate formally in the military discussion of the organization. The French navy in the Mediterranean, however, regularly takes part in joint exercises with the Sixth Fleet. In a much more substantive area, France and the United States probably have private understandings about each other's conduct in a nuclear war scenario so that their forces will not get in each other's way.

There appear to be several areas of common American-French interest. First, neither country wants to see the Soviet Union achieve the superior position in the Mediterranean. France's concern about the increased Soviet naval activity in the area was a major reason for the strengthening of its Mediterranean fleet.[33] Second, both the United States and France recognize the necessity for the West to have access to Arab and other Middle East oil at a reasonable price. Third, both countries have a vital interest in keeping open the sea lines of communication in the Mediterranean and realize the value of having a credible naval presence to accomplish this goal. France, additionally, has the incentive of using the Suez Canal to reach its garrison at Djibouti, in Afars and Issas Territory (French Somaliland) located on the Gulf of Aden at the southern entrance of the Red Sea and the gateway to the Indian Ocean. The United States, in my judgment, should encourage the build-up of another Western navy in the Indian Ocean to help counter the Soviet naval presence there. Last, there is the mutual interest in resolving the Cyprus dispute and the Arab-Israeli conflict.

---

[32] *Le Monde*, 21 November 1974.
[33] *Economist*, 10 May 1975.

The differences between France and the United States are also substantial—primarily competition for arms sales and foreign trade and rivalries over which country will be the leading spokesman for the West in the Mediterranean. France cannot realistically attempt to compete with the United States in all areas. In terms of economic strength and technology France is no match for the United States. But it is equally unrealistic for the United States to ignore France's ambition and ability to make a contribution. To do so almost guarantees France's behaving as a third force in the Mediterranean A far better course for both countries would be to work out a modus vivendi for a sharing of roles in the Mediterranean. The two countries certainly should not permit their rivalry to inhibit effective cooperation against the Soviet Union.

Great Britain. If its economy were stronger and if it could afford a larger defense establishment, Great Britain would be able to perform a more substantial role in the Mediterranean. However, a wide variety of economic problems—balance-of-payments difficulties, the rise of oil prices, a high rate of inflation, and a slow growth rate— have forced Britain to search for ways to save money. The main casualty of this effort to conserve funds has been the military, and the region most affected by this latest cut has been the Mediterranean. The British government announced plans to reduce its forces there this way:

> The Government has decided that it cannot in future commit British maritime forces to the Mediterranean in support of NATO. After 1976 no destroyers, frigates or coastal minesweepers will be earmarked for assignment to NATO in the area; and between 1977 and 1979 the Royal Air Force Nimrods and Canberras at present committed to NATO there will be withdrawn. In peacetime HM Ships will however visit the Mediterranean from time to time and will continue to participate in exercises there with our NATO Allies.
>
> The Government of Malta has been informed that we propose to arrange for the run-down of the British forces in Malta between 1 April 1977 and 31 March 1979 when the Military Facilities Agreement expires.[34]

Even with the rundown of forces west of Suez, Britain makes— and may continue to make—a significant contribution to Western security interests in the Mediterranean. The British presence in Malta

---

[34] *Statement on the Defence Estimates 1975*, p. 13.

Table 8

BRITISH NAVY VISITS IN THE MEDITERRANEAN,
1964 AND 1974

1964	1974
Algeria	Cyprus
Egypt	Egypt
France	France
Gibraltar	Gibraltar
Greece	Greece
Italy	Italy
Lebanon	Malta
Libya	Morocco
Malta	Tunisia
Morocco	Turkey
Spain	Yugoslavia
Tunisia	
Turkey	
Yugoslavia	

Source: Royal Navy.

and especially the possession of the two Sovereign Base Areas in Cyprus, at Akrotiri and Dhekelia, have been of positive value to the West. This survey of British forces in the Mediterranean will detail the scope of their presence in mid-1975.

Gibraltar. Ground troops consist of one infantry battalion and one detachment of Royal Marines. One detachment of vintage Hunter jet fighters is based on the island. A frigate makes Gibraltar its home port. A residual force will remain in Gibraltar.

Malta. By the end of 1979, all British forces are scheduled to be withdrawn from Malta, ending a presence of nearly two centuries. This evacuation entails the removal of Britain's Mediterranean headquarters, commanded by a rear admiral, one Nimrod and one Canberra reconnaissance squadron, and an 800-man Royal Marine commando battalion.

Cyprus. Britain will retain the two Sovereign Base Areas but will no longer permanently station aircraft there. Instead it will periodically rotate fighter units based in Britain to the island. By early 1975 some thinning of the air units had begun. The previous strength had been: two Vulcan bomber squadrons, which were assigned to the Central Treaty Organization (CENTO), one Lightning squadron, one C-130 Hercules squadron, and two Royal Air Force regiments, one equipped with Bloodhound surface-to-surface missiles. The Bloodhound SAM regiment will remain to provide protection for the airfields. Ground security will be maintained by an infantry ball battalion and an armed reconnaissance squadron, though these forces, too, are scheduled for reduction. In addition, another infantry battalion and an armed reconnaissance squadron are assigned to the United Nations peace-keeping force in Cyprus. As the conflict cools, Britain intends to cut these contingents as well. As a member of NATO, Britain also operates electronic monitoring installations in Cyprus and Turkey.

Britain still has considerable influence in the Mediterranean despite the humiliation it suffered, along with France, in the unsuccessful 1956 Suez Canal invasion. It also has the longest continuous presence of a non-Mediterranean power in the sea. Britain has brisk arms sales in the area, including an agreement concluded in June 1975 to provide about $1 billion worth of military hardware to Egypt.[35] As a power Britain can be expected to continue to play an active role within the region even though it is substantially reducing its forces there.

Additional Power Considerations. In the near future it is difficult to see any Mediterranean country other than France that is capable of exerting a significant influence throughout the entire region. Any number of countries could emerge as major regional powers, particularly if they acquired nuclear weapons. There are several likely candidates for joining the nuclear club already. In an April 1975 article in *Scientific American*, William Epstein, formerly director of the Disarmament Division of the United Nations Secretariat, identifies six Mediterranean countries now capable of acquiring nuclear weapons—though the required development periods differ.[36] According to Epstein, Italy and Israel could make such weapons "in a comparatively short time, given the political decision to do so."[37] Spain also could have nuclear arms in a relatively short time, Epstein says, while Algeria, Egypt, and Yugoslavia would require considerable time. These judgments, however, are based on the assumption that

---

35 *Washington Post*, 13 June 1975.
36 William Epstein, "The Proliferation of Nuclear Weapons," *Scientific American*, vol. 232, no. 4 (April 1975), pp. 18-33.
37 Ibid.

the weapons would be developed using domestic talent and techniques. It is conceivable that a country might be able to purchase nuclear arms or be supplied with such weapons under a regional defense pact. Libya reportedly approached China with such a proposition and was rebuffed.[38] Among other alarming possibilities, a terrorist group could steal nuclear weapons or highly radioactive and lethal nuclear waste material that could be packed into a conventional missile or bomb. There are both research and power reactors capable of supplying weapons-grade or radioactively lethal materials that could be diverted, in all Mediterranean countries except Morocco, Algeria, Tunisia, Libya, Malta, and Cyprus. Libya may acquire a reactor from the Soviet Union,[39] and Algeria, which places heavy emphasis on industrialization, may also look to atomic energy as a source for power and desalinization.

With the Suez Canal open, additional non-Mediterranean countries may develop political-military interests in the basin. Iran, for example, has invested in maritime, industrial, and housing projects in Port Said and sent two destroyers through the canal in the first north-bound convoy of ships that completed the reopening ceremonies.[40] China is another power that may augment its diplomatic and economic links with Mediterranean countries, such as Albania, by sending its navy on a Mediterranean cruise. Such an event would certainly receive wide press coverage and spark speculation about China's intentions. India, which is closer to the Mediterranean, may also wish to stretch its interests to include the region.

In this chapter I have attempted to construct a comprehensive list of the issues and problems that the United States is likely to encounter in the Mediterranean in the near future. Of course, many of these issues and problems could be dropped from the list and others could be added. The Mediterranean is an extremely unpredictable region. To complete this study, in my judgment, it is necessary to discuss how the United States can better understand and manage its foreign policy in the Mediterranean.

---

[38] *Economist,* 7 June 1975.
[39] *Washington Star,* 3 June 1975.
[40] *New York Times,* 7 June 1975.

6

TOWARD A MEDITERRANEAN CONSCIOUSNESS

A Mediterranean consciousness is emerging in America, but so far it is a vague, undefined notion that has not taken hold as a working policy concept. President Ford used the phrase "Atlantic and Mediterranean regions" during his May 1975 visit to Spain when discussing that country's contribution to Western security.[1] Secretary of Defense James R. Schlesinger talked about the "Mediterranean Basin" as an area of American interest and concern in a lengthy interview with the *New York Times*.[2] But as a bureaucratic operating framework for monitoring and assessing events or for executing American international security policy, the Mediterranean is obscured by the traditional national, regional, and land-oriented divisions recognized by the Departments of State and Defense.

How can this policy myopia be overcome? The answer, in my judgment, is very simple and requires only minor alterations in the present structure. I am not recommending carving up the existing geographical bureaus—taking the countries along the northern littoral away from Europe and those along the eastern and southern littorals away from the Near East and putting them in a Bureau for Mediterranean Affairs. Rather, I propose the assignment of fairly senior "Mediterranean watch officers" on an operating level in the State Department. Also, the State Department's Policy Planning Staff, Bureau of Intelligence and Research, and the Office of Political-Military Affairs, as well as the Defense Department's Office of International Security Affairs and the National Security Council at the White House, ought to include senior staff members who would constantly watch the Mediterranean area and frame policy options in

Washington Post, 1 June 1975.
New York Times, 5 May 1975.

the regional context. Other government agencies, the details of who organizational structures are not as apparent, such as the Centr Intelligence Agency and the Defense Intelligence Agency, ought consider whether they pay close enough attention to the Medite ranean Basin as a region.

It is not enough to review the strategic situation in the Medite ranean, as we now do, only during a crisis like the near showdov over the British bases in Malta in 1971–72. The Mediterranean is f too important an area to receive just occasional scrutiny. It requir a permanent apparatus that would continuously follow politic military, and economic developments in the region and assess the effects on American security interests.

One way to accomplish this would be to set up a standing wor ing group consisting of Mediterranean specialists from the Depar ments of State and Defense (including representatives from tl military services), the Central Intelligence Agency, the Nation Security Agency, and the National Security Council that would me regularly and frequently to review the Mediterranean. One of i specific tasks would be to monitor the activities of the Sixth Fleet ar the Fifth Escadra. The group could keep a running score sheet (port visits of both fleets, analysis of press coverage, and public rea tion to the visits, which in turn would help determine patterns visits, trends in public opinion, and relationships between the factors, and it could use the data it accumulated to make Americ policy proposals. This exercise could lead to more creative use naval power, not only in the Mediterranean but in other critic areas. It might also result in a greater appreciation of naval pow by diplomats and a better understanding of diplomacy and wor affairs by naval and military officers.

The Congress, particularly the committees whose responsibiliti are foreign affairs and the armed forces, could play a seminal role I encouraging an inter-regional approach to the Mediterranean.

Nongovernmental institutions, such as universities and priva research centers, can also play a role in developing this embryon awareness of the Mediterranean in the United States. Perhaps th is an easier and quicker way to begin to systematically consider tl Mediterranean as a region, given administrative inertia and tl reluctance of government departments to reorganize themselves. few starts have been made in the private sector. This study, f example, is the result of a grant from a private research center. Tl American University, in Washington, D. C., offers a course on tl Mediterranean as a zone of conflict in regional international relatior

A group of professors there have formed a Center for Mediterranean Studies that they hope will attract enough resources to allow them to conduct independent research and publish a journal. The American Universities Field Staff, which is operated by a consortium of 11 American universities, supports the Center for Mediterranean Studies in Rome. The Center for Strategic and International Studies of Georgetown University sponsored a conference in September 1972 on security problems in the western Mediterranean and later published the papers delivered at the meeting.[3] There may be other such efforts in the United States, but most universities' area studies programs parallel the traditional geographic divisions of the State Department.

Mediterranean Outlook

The dramatic shift in Egyptian policy toward the United States has far-reaching implications for the U.S. Sixth Fleet in the Mediterranean Sea—particularly in the eastern Mediterranean Basin. Egypt is considered the key to the Arab world. The widely publicized assistance that the U.S. Navy gave to Egypt to clear the Suez Canal could be an important catalyst in changing the generally negative Arab attitudes toward the Sixth Fleet that have existed since 1967. And the canal mine-clearing operation could turn out to be the watershed in reversing the political gains that the Soviet Union, with the support of the Fifth Escadra, has made in the Mediterranean. The presence of the Sixth Fleet flagship in the opening procession through the canal is dramatic evidence of this trend. Though the Sixth Fleet has the overwhelming preponderance of naval power in the Mediterranean Sea, the Soviet Union has skillfully used its fleet to enhance its national interests in the Mediterranean at the expense of the United States. It is as if the Soviet Union applied the principles of Mahan to the contemporary Mediterranean political scene. Surely, the United States can—and ought to—apply Mahan's theories of naval strategy in today's Mediterranean more successfully than the Soviet Union.

In the context of improved relations with Egypt, what other opportunities are there for the U.S. Navy to improve its image in the Arab world? How might they be used and what could they lead to? Expanded port visits to Egypt? More frequent visits to other countries? Could it reduce Soviet access? Could it have an echo in the

Alvin J. Cottrell and James D. Theberge, eds., *The Western Mediterranean: Its Political, Economic and Strategic Importance* (New York: Praeger, 1974).

Persian Gulf? The Indian Ocean? Are there opportunities for nav weapons systems aid (including ships and/or training) to Arab cou tries? The answers to these and many other related questions a now of crucial importance to the U.S. Navy and the U.S. governme as a whole. The entry into service of Russia's first fixed-wing aircra carrier adds a sense of urgency to a full, candid, and regular di cussion of these questions. One intriguing question, for exampl is whether there will be a reduction in the nervousness of our NAT allies in the Mediterranean about identifying themselves with U. policy interests in that area if there is substantial, across-the-boar improvement in American-Arab relations.

Agenda for Future Study. There are scores of other questions who answers bear on American interests:

(1) How is the change in the balance of naval power viewed the Arab world? Is the presence of the fleets of the two supe powers seen as a stabilizing factor in international relations the Mediterranean, particularly in the Middle East, or as irritant? Does it make any difference?

(2) How is the presence of American and Soviet naval pow in the Mediterranean perceived by the littoral countries? A the perceptions of political leaders different from those of th man-in-the-street?

(3) Are naval activities regularly covered in the Arab media– both press and radio-television? As news items? Subjects fo commentary? For example, was there coverage of such develop ments as the U.S. Sixth Fleet homeporting in Greece; the statior ing of an American submarine tender in Sardinia; the first Sovie aircraft carrier; the Soviet fleet using Latakia and Alexandri as naval bases? Before June 1967, the U.S. Sixth Fleet mad visits from time to time to Beirut. How were these visits viewed What would be the reaction if a Soviet navy ship called at Beirut What would be the reaction if the U.S. Navy resumed its visits

(4) Recently several suggestions by Mediterranean countrie have been made about controlling pollution and regulating othe affairs of the sea. Several meetings were held to try to prepare Mediterranean position for the Law of the Sea Conference. has been proposed that the fleets of the superpowers be expelle Is the fact that 17 nations share the Mediterranean a basis fo cooperation? What kind—economic, pollution control, politica defense? Ought such cooperation be encouraged by the Unite States? In any event, it is clear that more Mediterranean cou

tries will be increasingly assertive about activities near their shores. These activities include naval and shipping matters as well as mineral exploration and intelligence-gathering operations. On 21 March 1973, for example, two Libyan Mirage fighter jets fired shots at, but missed, an American EC-130 (the electronic intelligence version of the Hercules transport) while it was cruising 83 miles from Libya's coastline.[4] Libya had declared a 100-mile "restricted air zone" that foreign aircraft could not enter without permission [5]—a claim that the United States was aware of but rejected. Among the many running disputes between Greece and Turkey is a quarrel over oil exploration rights in the eastern Aegean Sea that some observers feel could be more volatile than their differences over Cyprus.

Conclusion

The maritime life of the Mediterranean has been developing since man built his first cities along the coastline and first put his boats to sea. For at least 6,000 years, people, goods, and ideas have criss-crossed this vital area. It is rich in mythology, legend, and history: The lore of the Mediterranean embraces Jason and Ulysses; Moses, Jesus, and Muhammad; Socrates and Aristotle; Tutankhamun, Xerxes, Pericles, Hannibal, Caesar, Barbarossa, and Andrea Doria; Sparta, Rome, Carthage, and Constantinople; the Battle of Salamis, the Cru-sades, Lepanto, and the Battle of the Nile; Napoleon and Nelson.

It wasn't too long ago that Greek and Italian were the languages of commerce in Port Said and Alexandria. Spanish is still the tongue most frequently heard in Tangier. However, the most pervasive and lasting influence in today's eastern and southern Mediterranean is the Arab conquest in the late seventh and early eighth centuries when the Arabs gushed out of the Arabian Peninsula carrying the standard of Islam, swept along the southern shore of the "Sea of the Romans," as they called it then, across to Spain and into France until they were defeated by Charles Martel at Poitiers in 732. The Arabs eventually withdrew from Europe leaving a legacy of architecture and place names. Gibraltar, for example, is named after Tarik, the Berber soldier who made the first raid on the Iberian Peninsula in 711. In the west, Jebel Tarik (for Tarik's mountain) has been corrupted to

---

[4] *New York Times*, 22 March 1973. The plane was part of a U.S. Air Force unit based at the American facility at Athens International Airport, which helps explain official Greek concern about the American presence.

[5] Ibid., 24 March 1975.

Gibraltar, but it is still the Arabic name for The Rock. Arabic remains the dominant language of the Middle East and North Africa, and Islam is practiced by their inhabitants. The Maltese, though thoroughly Roman Catholic, speak an Arabic dialect heavily laced with Italian and Spanish, but Arabic enough to enable them to understand broadcasts from Libya's Radio Tripoli just 100 miles away. And, of course, today Arab oil politics affect the entire basin and most of the rest of the world.

The northern Mediterranean is more of a mosaic, reflecting a more complex history. Its inhabitants are Spaniards, Basques, Frenchmen, Italians, Corsicans, Sicilians, Serbs, Croats, Greeks, Turks, and Armenians. And in Israel, Jews from four continents representing many backgrounds and languages blend into a new nation. The Mediterranean is also a melting pot of religions. There are Roman Catholic, Maronite, Copt, Syriac, and Orthodox versions of Christianity; Sunni, Shia, Druze, and Alawite Islam; Hasidic, Orthodox, and Reform Judaism.

There are few new cities in the Mediterranean. A few hundred yards from the present thriving port of Piraeus is the ancient harbor of Piraeus that is today the fashionable Marina Zea for private yachts. Somewhere under centuries of silt and more recent asphalt are the ruins of the city that was founded by Alexander the Great.

More recently, the first foreign adventure of the American navy was to force respect for the infant American flag along North Africa's Barbary Coast. This deployment in the late eighteenth century was a forerunner of today's Sixth Fleet. The men who fought there— Preble, Truxton, Decatur, Bainbridge—are legendary in American naval history. The naval battle off Cape Trafalgar on 21 October 1805, during which Nelson decisively interrupted Napoleon's plans to dominate Europe, was both the end and beginning of an era. Trafalgar was the last great set piece naval engagement under sail, and it began a period of British naval supremacy that was to last until World War II.

Ironically, 162 years to the day after Nelson defeated the combined French and Spanish fleets near the western entrance to the Mediterranean, at the other end of the sea a Russian-built Styx missile fired by an Egyptian crew sank the Israeli destroyer *Elath*. Both events are turning points in naval warfare. The twenty-first of October 1967 was the first time in naval history that a surface-launched antishipping missile had been fired in anger. The *Elath* incident—a sort of naval sputnik—demonstrated the effectiveness of a new weapons system that has since been widely adopted by Western navies.

All of this history, ancient, medieval, and modern, has left its indelible stamp on the economics and politics of today's Mediterranean. The Italian colonization of Libya is still a political fact of life, so much so that the regime has had the bodies of Italians buried in Libya exhumed and returned to Italy. French relations with Algeria are still complicated by their colonial history. Yet, French is spoken and read by the majority of Algerians and the largest daily newspaper in Algiers, *Al Moujdahid*, is printed in French. Two notions of history, one ancient, the other medieval, clash in today's Middle East. The dispersal of the Jews and their burning desire to re-create and maintain their homeland run violently against the nationalist grain of the Arab world, which is the core reason for the present conflict.

The lessons to be learned from the Mediterranean's history and contemporary developments are myriad. In many ways this region is a microcosm of today's world. It is my hope that this book will help develop a Mediterranean consciousness in the United States.

APPENDIX A
WEAPONS IN THE MEDITERRANEAN

NAVAL STRENGTHS OF MEDITERRANEAN COUNTRIES

Country/Vessel Type	Number
Albania	
Submarines	4
Escorts	4
Motor torpedo boats	42
Patrol boats	10
Mine countermeasure ships	8
Total	68
Algeria	
Submarine chasers	6
Osa-class missile boats [a]	3
Komar-class missile boats [a]	6
Motor torpedo boats	12
Fleet minesweepers	2
Total	29
Cyprus	
Egypt	
Submarines	12
Destroyers	5
Escorts	3
Submarine chasers	12
Osa-class missile boats [a]	8
Komar-class missile boats [a]	5
Motor torpedo boats	30
Minesweepers	12
Landing craft	14
Total	101
France	
Aircraft carriers	2
Submarines	22
(3 SSBN)	
(2 SSBN under construction)	
(19 attack, diesel powered)	
(4 more under construction)	
Cruisers	2
(1 with SSM and SAM)	
(1 with heavy ASW helicopters)	
Destroyers	19
(2 more in service in 1975)	
(2 with SAM and ASW)	
(6 with SAM)	
(7 with ASW)	
Frigates	24
(3 more in service in 1975)	
Patrol boats	27
(1 with SSM)	

Country/Vessel Type	Number
Fleet minesweepers	8
Coastal minesweeper/hunters	38
Landing craft	22
Total	164
Greece	
Submarines	7
Destroyers	11
Escorts	4
Missile boats [b]	4
(4 more on order)	
Motor torpedo boats	12
Patrol boats	5
Minesweepers	15
Minelayers	2
Landing craft	22
Total	82
Israel	
Submarines	2
(3 more on order)	
Reshef-class missile boats [c]	6
Saar-class missile boats [c]	12
Motor torpedo boats	6
Patrol boats	30
Landing craft	10
Total	66
Italy	
Submarines	10
(2 more under construction)	
Cruisers	3
(2 with SAM and ASW helicopters)	
(1 with ASW helicopters and ASROC)	
Destroyers	9
(4 with ASW helicopters and SAM)	
Frigates	18
(6 with ASW helicopters)	
Patrol boats	10
(2 hydrofoils with Otomat SSM)	
(2 with Seakiller Mk2 SSM)	
Minesweepers	55
(4 ocean)	
(31 coastal)	
(20 inshore)	
Landing craft	66
Total	171
Lebanon	
Patrol boats	5
(3 coastal patrol boats on order)	
Landing craft	1
Total	6

Table A-1 (continued)

Country/Vessel Type	Number
Libya	
Frigates (with Seacat SAM)	1
Corvettes	1
Fast patrol boats	3
(each with 8 SS-12M SSM)	
(14 more on order)	
Patrol craft	11
Support/supply ships	1
Total	17
Malta	
Morocco	
Frigates	1
Coastal escorts	2
Patrol boats	1
(2 more on order)	
Landing craft	1
Total	5
Spain	
Helicopter carriers[d]	1
Submarines	10
Cruisers	1
Destroyers	13
Frigates	10
(1 more on order)	
(2 with SAM and ASROC)	
Corvettes	4
Torpedo boats	2
Minesweepers	18
Landing craft	16
Patrol boats	18
Total	93
Syria	
Komar- and Osa-class SSM missile boats[a]	6
Torpedo boats	11
Patrol boats	1
Minesweepers	1
Total	19
Tunisia	
Destroyer escorts	1
Corvettes	1
Patrol boats	15
(2 with SS-12M SSM)	
(1 more on order)	
Minesweepers	1
Total	18

Table A-1 (continued)

Country/Vessel Type	Number
Turkey	
Submarines	16
(1 more under construction)	
Destroyers	13
Escorts	5
Patrol boats	70
Minesweepers	20
Minelayers	9
Landing craft	50
Missile boats	—
(4 on order)	
Total	183
Yugoslavia	
Submarines	5
Destroyers	1
Osa-class missile boats [a]	10
Torpedo boats	34
Patrol boats	26
Minesweepers	30
Landing craft	31
Corvettes	3
Total	140

[a] Armed with Russian "Styx" surface-to-surface missiles (SSM).

[b] Armed with French-built Exocet surface-to-surface missiles.

[c] Armed with Gabriel surface missiles (SSM) of Israeli design and manufacture

[d] The Spanish navy has ordered 8 Harrier V/STOL aircraft that are suited fo operations from this helicopter carrier.

Source: *The Military Balance 1975–76.*

Table A-2

NAVAL SURFACE-TO-SURFACE MISSILES
IN THE MEDITERRANEAN

Country of Possession	Type
Algeria	Styx (on 6 Komar- and 3 Osa-class fast patrol boats)
Egypt	Styx (on 8 Osa- and 5 Komar- class fast patrol boats)
France	Exocet (on 1 cruiser and 2 destroyers) SS-11 (on 1 patrol craft)
Greece	Exocet (on 4 fast patrol boats)
Israel	Gabriel (on 6 Reshef-class fast patrol boats) (on 12 Saar-class fast patrol boats)
Italy	Seakiller Mk2 (on 2 fast patrol boats) Otomat (on 2 hydrofoils)
Libya	SS-12M (on 3 fast patrol boats)
Syria	Styx (on 3 Komar- and 3 Osa-class fast patrol boats)
Tunisia	SS-12M (on 2 patrol boats—1 on order)
Yugoslavia	Styx (on 10 Osa-class fast patrol boats)

Source: *The Military Balance 1975–76.*

Table A-3

CHARACTERISTICS OF NAVAL SURFACE-TO-SURFACE AND SUB-SURFACE MISSILES

Missile	Range (nautical miles)	Characteristics
Surface-to-surface		
Otomat	43–108	This guided missile is powered by a turbojet though assisted at launching by two jettisonable booster rockets. It flies the last two miles to target 50 feet above sea level using a radio altimeter. It is used by the Italian navy on its "Freccia"-class fast patrol boats.
Exocet	20	Designed to provide warships with an attack capability against other surface ships. The system is compact enough to be fitted on patrol boats as well as larger vessels. It is powered by a two-stage solid-fuel rocket and can be effective in an ECM environment. Immediately after launch it flies about 10 feet above sea level using a radio altimeter and presents a difficult target to defenders. The speed is described by *Jane's* as "high subsonic."
SS-11	1.6	Developed by the French for its navy and army, the SS-11 is a two-stage rocket that can be fitted with a wide range of warheads—anti-tank, anti-personnel and perforating/exploding. Guidance is visual/manual by wire.
SS-12	3.2	A larger and more powerful version of the SS-11 and is an exclusively naval weapon. It is used by three Libyan navy patrol boats in addition to France.
Gabriel	14–26	Operated by the Saar- and Saar IV-class patrol boats of the Israeli navy. It is guided by radar but can also be guided visually to overcome ECM. It, too, travels a few feet above sea level and can be used in rough weather. An improved version has a 26-mile range.
Sea Killer (Mk 1)	1.6/5.4	Also called the Nettuno, this single-stage solid-propellant rocket has a minimum and maximum range for greater flexibility. It is also a surface skimmer and it has a high-explosive warhead with a proximity fuse. Operational in the Italian navy.
Sea Killer (Mk 2)	13+	Dubbed the Vulcano, this is a two-stage version of the Mk 1 to improve its range and increase the punch of its warhead. Operational in the Italian navy.
Sea Killer (Mk 3)	24+	A further improvement of this missile currently being developed.
Harpoon	60	A U.S. radar-guided missile under development. It may be launched from surface ships, submarines and planes, and is designed for increased offensive capability against surface ships.

(SS-N-2)		...missile is believed to have an active radar homing capability and automatic/command guidance system. Operational in 1960. Proved its effectiveness when Egyptian missile boats sank Israeli destroyer Elath in October 1967.
Shaddock (SS-N-3)	243	Largest of Soviet naval cruise missiles. Carried in cannisters and launched by jettisonable booster rockets, the Shaddock has a command guidance system that can receive target information in mid-course by other vessels or aircraft. Terminal guidance is by its own infrared homing. This weapon system, which can carry conventional or nuclear warheads, is widely deployed in the Mediterranean on Soviet surface ships and submarines. It is not believed to be possessed by any other navy.
SS-N-7	26	A Styx type of Soviet naval cruise missile that may also have an underwater launching capability.
SS-N-9	150	Little is known about this weapon except its range is thought to be up to 150 nautical miles with the normal range being about 40 nautical miles. It is carried on Nanutchka-class corvettes and possibly on J- and P-class submarines.
SS-N-10	30	Carried on Kresta II and Kara cruisers and on Krivak destroyers.
SS-N-11	20 [a]	Carried on Osa II patrol boats and on modified Kildin destroyers.
Samlet	108	A jet engine cruise missile. This is the surface-to-surface version of the Soviet air-to-surface missile, Kennel.
Sub-surface		
Malafon	7	This antisubmarine weapon is launched by a surface ship with jettisonable rockets and glides to the general target area where the torpedo separates from the parent vehicle and enters the water and behaves like a conventional homing torpedo. It is a standard ASW weapon of the French navy.
Asroc	1 to 5	Asroc (for antisubmarine rocket) is in wide service in the U.S. Navy and is regularly deployed in the Mediterranean. The system operates basically the same as the Malafon except it can also be used as a depth charge in addition to a homing torpedo.
Subroc	30	Subroc (for submarine rocket) is part of a complex ASW weapon system using submarines, aircraft and surface vessels. The missile is launched from a submerged submarine and is guided to the patch of water above the target. The bomb, or depth charge, then sinks and explodes. Can be fitted with a nuclear warhead.

[a] Estimated.

Source: *Jane's Weapon Systems 1974–75.*

133

Table A-4
MAJOR ADVANCED COMBAT AIRCRAFT
IN THE MEDITERRANEAN

Country/Aircraft	Type	Numb
Albania		
MiG-21/F8 [a]	Interceptor	1?
MiG-19/F6 [a]	Interceptor	3?
MiG-15 [a]	Fighter-bomber	2?
MiG-17	Fighter, ground attack	2?
Total		9?
Algeria		
MiG-21	Interceptor	3?
MiG-17	Fighter, ground attack	7?
MiG-15	Fighter, ground attack	1?
Su-7BM	Fighter, ground attack	2?
IL-28	Light bomber	2?
Magister	Ground attack, counter-insurgency	2?
Total		186
Egypt		
MiG-23	(48 more being delivered)	?
MiG-21	Interceptor (with Atoll AAM)	250
MiG-17	Fighter-bomber	125
Su-7	Fighter-bomber	80
Tu-16D/G	Medium bomber (10 with Kelt ASM)	25
IL-28	Light bomber	5
Mirage V	(from Saudi Arabia)	38
Mirage F-1	Interceptor (44 on order)	—
Total		523
France		
Mirage IVA	Strategic bombers (16 in reserve)	52
Mirage IIIC	Interceptor	45
Mirage F1	Interceptor	45
Super Mystere B2	Interceptor	45
Mirage IIIE	Fighter-bomber	120
Mirage VF	Fighter-bomber	30
F-100D	Fighter-bomber	56
Jaguar	Fighter-bomber	60
Vautour	Light bomber (being phased out)	15
Mirage IIIR/RD	Reconnaissance	45
Total		513
Greece		
F-4E	Fighter-bomber	36
F-104G	Fighter-bomber	20
F-84F	Fighter-bomber	62

Table A-4 (continued)

ountry/Aircraft	Type	Number
reece (continued)		
F-5A	Interceptor	72
F-102A	Interceptor	16
RF-5A	Reconnaissance	14
RF-84F	Reconnaissance	18
HU-16B	Maritime reconnaissance	12
A-7D	Tactical fighter (60 on order)	—
Mirage F1	Interceptor (40 on order)	—
Total		250
rael [c]		
F-4E Phantom	Fighter-bomber/interceptor (35 on order)	200
Mirage III/Kfir	Fighter-bomber/interceptor	75
A-4E/F/N Skyhawk	Fighter-bomber (20 on order)	200
Super Mystere B2	Interceptor (in reserve)	12
RF-4E	Reconnaissance	6
Mystere IVA	Fighter-bomber (in reserve)	23
Ouragan	Fighter-bomber (in storage)	30
Vautour	Light bomber (in storage)	10
Total		556
ly		
F-104S	All-weather fighter	164
F-104S	Fighter-bomber	18
F-104G	Fighter-bomber	36
G-91Y	Fighter-bomber	36
G-91R	Light attack bomber	35
RF-104G	Reconnaissance	30
Breguet 1150 Atlantic	Maritime reconnaissance	18
S-2 tracker	Maritime reconnaissance	20
PD-808 Vespa Jet	Electronic reconnaissance	15
Total		372
banon		
Mirage IIIEL	Interceptor	6
(with R-530 AAM)	(4 in storage)	4
Hunter	Fighter, ground attack	13
Mirage IIIBL	Interceptor (in storage)	1
Total		24
ya		
Mirage IIIE	Interceptor	32
Mirage V	Fighter, ground attack	50
Mirage IIIER	Reconnaissance	10

135

Country/Aircraft	Type	Num
Libya (continued)		
MiG-23	Fighter, ground attack (29 on order)	—
Tu-22	Bomber (12 on order)	—
Total		9:
Morocco		
F-5A/B	Interceptor	2
Magister	Fighter, ground attack	2
MiG-17	Fighter-bomber (in storage)	1:
Total		6(
Spain		
F-4C	Fighter-bomber	3(
Mirage IIIE/IIIDE	Fighter-bomber	3(
F-5A/B	Fighter-bomber/Reconnaissance	4(
HA-220 Super Saeta	Ground attack	2(
HA-200D Saeta	Ground attack	4!
HU-16B Albatross	Maritime reconnaissance	1
P-3	Maritime reconnaissance	
Mirage F1-C	Interceptor (15 on order)	—
Total		19
Syria		
MiG-23	Tactical fighter	4!
MiG-21	Interceptor (more on order)	25(
MiG-17	Fighter, ground attack	5(
Su-7	Fighter-bomber	4!
IL-28	Light bomber	
Total		39(
Tunisia		
F-86F	Fighter	1:
SF-260W Warrior	Counter-insurgency aircraft	1:
Total		24
Turkey		
F-4E	Fighter-bomber (more on order)	2C
F-104G	Fighter-bomber	33
F-100D	Fighter-bomber	4!
F-5A	Fighter-bomber	32
F-5A	Interceptor	16
F-102A	All-weather interceptor	36
RF-84F	Reconnaissance	20
RF-5A	Reconnaissance	4C
F-104S	Fighter-bomber (more on order)	18
F-84F	Fighter, ground attack	32
Total		292

Table A-4 (continued)

Country/Aircraft	Type	Number
Yugoslavia		
MiG-21	Fighter-bomber	110
Galeb/Jastreb	Fighter, ground attack	95
Kraguj	Fighter and ground attack	15
F-84	Fighter, ground attack	10
RT-33A	Reconnaissance	15
Galeb/Jastreb	Reconnaissance	25
Total		270

a Chinese-supplied.

b Egypt also has 200 training aircraft (including most of the types listed) that could be used during hostilities in extreme circumstances. The figures are approximate because of losses during the October 1973 war and subsequent replacements.

c In addition, Israel has 85 Magister trainers.

d Some Mirages may be in storage.

Source: *The Military Balance 1975–76.*

Table A-5

CHARACTERISTICS OF MAJOR ADVANCED AIRCRAFT IN THE MEDITERRANEAN

Aircraft Type	Ordnance (in pounds)	Range (nautical miles)	Avionics, Weaponry and Operational Role
MiG-23	Unknown	520 (combat radius)	Little is known about the avionics, but the aircraft was probably designed to intercept fast strike aircraft, possibly with "snap-down" missiles to deal with low-flying raiders. It presumably carries air-to-air guided weapons.
MiG-21	1,200	593 (internal fuel) 971 (ferry, with 3 external tanks)	The MiG-21 is fitted with search-and-track plus warning radar and Atoll air-to-air missiles with a probable infrared guidance system similar to the American-made Sidewinder 1A.
Su-11	Unknown	Unknown	The Su-11 was possibly developed to meet the Soviet requirement for a faster interceptor to replace the Su-9. Normally it carries one radar homing and one infrared homing Anab air-to-air missile. There are provisions for additional weapons or fuel.
Su-7	4,400	172-260 (combat radius) 780 (range)	Subsequent to 1961, the Su-7 became the standard tactical fighter-bomber of the Soviet air force. Little information is available concerning the avionics of the Su-7.
Mirage V	8,800	347-699 (combat radius) 2,158 (ferry, with 3 external tanks)	The Mirage V can carry one Matra R5 30 all-weather air-to-air missile with radar or infrared homing heads, an air-to-surface missile, and two Sidewinder air-to-air missiles. Avionics include air-to-air capabilities for cannons and missiles and an air-to-ground capability for dive bombing.
F-4E Interceptors Ground attack	16,000	1,997 (ferry) 781+ (combat radius) 868+ (combat radius)	The F-4E uses highly sophisticated electronic countermeasures equipment, computers, and radar in its role as a long-range all-weather attack fighter. It is considered to be the best American missile-armed aircraft.

Aircraft		Range	Characteristics
A-4E	10,000	1,785 (ferry)	The A-4E fighter-bomber is equipped with an angle-of-attack indicator, terrain-clearance radar, and a variety of optional sophisticated weapons such as air-to-air and air-to-surface rockets (Sidewinders), infrared Bullpup, air-to-surface missiles and torpedoes).
Mirage IIIC	2,000	647 (combat radius for ground attack version)	The characteristics of the Mirage IIIC are the same as for the Mirage V, except that the planes supplied to Israel are said to have a different electronic configuration. Israel has developed an infrared homing air-to-air missile with a "see-and-shoot" capability and has fitted them to its Mirage fighters.
Ilyushin IL-28	4,400	2,500 (maximum with internal fuel)	IL-28 carries two fixed forward-firing automatic aircraft guns thought to be 23mm. NS type, two 23mm. guns in tail turret, radar for navigation, and blind bombing. Utilized as a standard Soviet tactical bomber for many years.
MiG-19		739 (maximum with external tanks) 522 (normal)	Can carry one 37mm. and two 23mm. cannons, or two 23mm. cannons, or three 23mm. cannons. Can also deliver air-to-air rockets or total of 1,100 pounds of bombs.
Aeritalia G91Y	4,000	1,890 (ferry, with maximum fuel) 323 (typical combat radius)	Lightweight fighter-bomber and reconnaissance aircraft. It has a 60 percent greater takeoff thrust than the G91 so there is a greater possible military load.
Hunter	8,140	1,600 (no reserve)	(MK9): Four 30mm. Adenguns (150 rpg). Single-seat fighter aircraft which possesses automatic gun ranging radar with scanner and gyro gunsight and which can carry a variety of rockets of other weapons.
F1-C	8,800	Unknown	The F1 can carry Matra R.530 or Super 530 radar homing or infrared homing air-to-air missiles. Stores can also include a Martel anti-radar missile or an AS.30 air-to-surface missile. Avionics can include fire-control radar, bombing computer, laser rangefinder and remote-setting interception system.

Source: *Jane's All the World's Aircraft 1974–75.*

Table A-6

CHARACTERISTICS OF MAJOR AIR-TO-SURFACE MISSILES

Missile	Range (nautical miles)	Characteristics
AS-37/AJ-168 Martel	32	An air-to-surface tactical guided missile that comes in two versions. One (the AS-37) is an anti-radar model with a passive radar homing guidance system. The other (AJ-168) has a visual command guidance system with a TV camera and data link between the missile and aircraft. Both versions carry a proximity fused warhead and can operate effectively in an electronic countermeasures environment. Martel can be carried by a variety of aircraft, including Hawker-Harrier, Mirage IIIE, Jaguar, Atlantic, and Buccaneer Mk11.
Albatross	32-43	This anti-ship missile is the air-launched version of the Otomat anti-ship SSM. This missile can also be carried by a variety of aircraft such as Atlantic, Super Etendard, Nimrod, and the helicopters SH3D and Super Frelon.
HARM		HARM (High-speed anti-radiation missile) is being designed as an improvement over the existing Shrike and Standard missiles. To be used in defense suppression operations. HARM will have greater speed, faster reaction and a more destructive warhead than existing missiles.
Harpoon (AGM-84A)	60	This is the air-launched version of the Harpoon anti-ship SSM. It has been launched from the P-3 Orion aircraft.
Bulldog (AGM-83A)		This is a laser-guided, air-to-surface tactical missile designed for close air support. Capable of being fired at long range and low altitudes, Bulldog relies on a Forward Air Controller, either land-based or airborne, which selects targets, transmits instructions to the attack aircraft, operates the laser designator, and illuminates the selected target. The missile finds its target automatically by its laser-seeking homing head, and carries a Mk19 Med 0 warhead. It can be carried by the A-4, A-6, and A-7 aircraft.
Bullpup (AGM-12)	6-9	An air-to-surface missile used for attacking tactical surface targets on land or sea. Bullpup A has an optical/command guidance system with a radar altimeter guidance system for its warhead which may be either a hollow charge or fragmentation type. Bullpup B has an automatic/command guidance system with a laser-guided warhead

which may be either high-explosive or nuclear. Its maximum speed is over Mach 2. Bullpup may be carried by the A-4 Skyhawk, P-3 Orion, F-4 Phantom, A-5 Vigilante, A-6 Intruder, F-8E Crusader, F-100, F-105, and the British Sea Vixen and Buccaneer aircraft.

Condor (AGM-53A) — 32-43

Condor is a medium-range supersonic air-to-surface cruise missile used in stand-off capacities against heavily defended ground targets. It has been developed to replace Bullpup missiles. Condor uses a TV remote guidance and control system and carries a conventional high-explosive warhead. A newer version is being developed with dual mode radar and electro-optical seeker that will permit a night/all-weather capability not possessed by the current version. Carried by the A-6 Intruder, A-7 Corsair II, and will be carried by the F-14.

AS-12 — In relation to surface —26,250 feet; to aircraft—18,000 feet

The AS-12 is a lightweight wire-guided air-to-surface general purpose missile. An automatic guidance system is being developed for newer models. AS-12 may be carried by various aircraft and helicopters and may carry different types of warheads weighing up to 62 pounds. It may be used against fortifications, tanks, ships, and other vehicles.

AS-30 — 5.4-6.4

The AS-30 is a medium/heavy air-to-surface tactical missile. It has an optical/automatic/command guidance system and carries a high-explosive, infrared guided warhead, with alternative delay or non-delay fuses. Minimum launching speed is Mach 0.45 and thus AS-30 can be carried by any aircraft capable of launching at such speed.

AS-30L

This is a lighter version of the AS-30 and performs the same tactical role. The warhead it carries is about half the size of the AS-30's.

Shrike (AGM-45A) — 5.4

A U.S.-made air-to-surface missile used against ground-based defensive radar. The missile contains a radar receiver which senses the direction of the radar and commands the guidance system toward it. The warhead has an active/passive homing radar system. Carried on F-105, F-4, A-4, A-6, and A-7.

Kennel (AS-1) — 63

This is the air-to-surface version of the Soviet naval SSM Samlet. Warhead has passive/active homing radar. Carried by Tu-16B Badger, a twin-jet long-range strategic bomber with a speed of 587 mph: believed designed for anti-shipping missions. (Badger/Kennel combination has been supplied to Egypt.)

Kipper (AS-2) — 155

Soviet anti-shipping stand-off air-to-surface missile, carried by Tu-16C Badger. Operation believed similar to Kennel missile. Also similar to U.S.-made Hound Dog. Missile carries nuclear warhead.

Table A-6 (continued)

Missile	Range (nautical miles)	Characteristics
Kangaroo (AS-3)	400	Largest air-to-surface missile in Soviet forces. Armed with a nuclear warhead, it is carried by the Tu-95B Bear (also called the Tu-20), a four-Turboprop long-range bomber with a speed of 500 mph. It is believed to have either an inertial or command guidance system.
Kitchen (AS-4)	400	This is believed to be the most advanced in the Soviet line of air-to-surface stand-off missiles. Carried by the Tu-22B Blinder, a twin-jet bomber with a speed of Mach 1.4, it has a nuclear warhead and an inertial guidance system.

Source: *Jane's Weapon Systems 1974–75.*

Table A-7

MAJOR SURFACE-TO-AIR GUIDED MISSILES AND RADAR-DIRECTED GUNS IN THE MEDITERRANEAN

Country of Possession	SAMs	Number
Albania	SA-2 SAM	X
Algeria		
Egypt	SA-2	360
	SA-3	200
	SA-6	75
	ZSU-23-4 SP AA guns	X
	ZSU-57-2 SP AA guns	X
	SA-7	X
France	Hawk	X
	Roland	X
	⎰ Mascura SAM	(on 3 destroyers)
	⎱ Tartar	(on 4 destroyers)
Greece	Hawk	12
	Nike Hercules	X
Israel	Hawk (more on order)	90
	Redeye	X
Italy	Hawk	X
	⎧ Terrier	(on 2 cruisers)
	⎨ Standard	(on 4 destroyers)
	⎩ Tartar	(on 4 destroyers)
	Nike Hercules	96
Lebanon		
Libya	Seacat SAM	(on 1 frigate)
	⎧ Crotale	60
	⎪ SA-2	X
	⎨ SA-3	X
	⎩ SA-6	X
Morocco		
Spain	Nike Hercules	(in 1 battalion)
	Hawk	(in 1 battalion)
	Standard	(on 2 frigates)
Syria	SA-2	(in 24 batteries)
	SA-6	(in 14 batteries)
	SA-3	(in 24 batteries)
	SA-7	X
	SA-9	X
Turkey	Nike-Ajax/Hercules	20
Yugoslavia	ZSU-57-2 SP AA guns	X
	SA-2	(in 8 batteries)

Note: SP AA = self-propelled antiaircraft.
X = included, but exact number not available.
Source: *The Military Balance 1975–76.*

Table A-8

CHARACTERISTICS OF SURFACE-TO-AIR MISSILES AND OTHER MAJOR ANTI-AIRCRAFT WEAPONS

Missile	Characteristics	Effective Ceiling
American		
Hawk	This is an American-made guided weapon with a solid propellant motor and a continuous wave, semi-active radar-homing system. It is said to be effective against aircraft flying at normal combat altitudes down to tree-top level.	Over 35,000 feet
Nike-Ajax	Less effective predecessor of Nike-Hercules.	Over about 27 miles
Nike-Hercules	This American guided missile system has been successful against high-performance aircraft at different altitudes as well as against short-range ballistic missiles. It has a command guidance system on both missile and warhead, plus a command detonation system for the warhead, which can be either nuclear or high-explosive. System includes acquisition radar, target tracking radar and remote control launchers. Each battery can operate alone or as part of a network even in an electronic counter-measures environment. Speed: supersonic. Range: more than 86 miles.	Over about 27 miles
Terrier	A U.S. Navy-developed SAM capable of carrying either a high-explosive or nuclear warhead. Its missile has a beam riding/semi-active homing radar with proximity fused warhead. Range: 21 miles. Speed: Mach 2.5.	Over 12 miles
Tartar	A U.S. SAM for primary defense of destroyers. Missile has semi-active homing radar with direct action and proximity-fused high-explosive warhead. Slant range: more than 10 miles. Speed: Mach 2.	—
Standard	A naval SAM used as replacement for Tartar and Terrier. Two models: one for medium range (M.R.), the other for extended range (E.R.) use against low-flying aircraft and anti-ship cruise missiles. Both versions (of which there are 8 variants) are equipped with high-explosive proximity fused warheads on missiles with semi-active homing radar. Speed: M.R.—over Mach 2; E.R.—over Mach 2.5. Range: M.R.—over 11 miles; E.R.—over 34 miles.	—

Name	Description	
Redeye	A portable shoulder-fired guided antiaircraft missile optically aimed with infrared homing guidance system. Redeye carries a high-explosive warhead and is designed for use against low-flying aircraft. Speed: supersonic. Range: probably 3 kilometers.	—
Talos	A naval SAM, later versions of which have a surface-to-surface capability. It has a beam-riding guidance system with a semi-active homing radar/proximity fused/continuous rod warhead of optional nuclear or high-explosive type. Range: 74 miles. Speed: Mach 2.5.	Over 84,000 feet
Canadian		
Sea Sparrow	A naval SAM/SSM with a semi-active homing radar guidance system and a proximity fused warhead. The system consists of launching and fire control systems which can provide early warning, simultaneous air and surface target tracking and simultaneous operation of missiles in single or rapid succession against air and surface targets. Slant range: 8 miles.	Unknown
Soviet		
SA-2	The SA-2 is a two-stage missile with automatic radio command. Target aircraft are tracked by radar which feeds signals to a computer, from which radio signals to the missile are generated. Those captured by the Israelis in 1967 had warheads detonated by contact or proximity fuses. American pilots in Vietnam have reported that the SA-2s there lack the maneuverability needed to intercept high-speed aircraft performing evasive tight turns.	60,000 feet
SA-3	The SA-3 is a two-stage, compact, mobile land-based missile used for short-range defense against aircraft at very low altitudes. It is similar to the American-made Hawk.	40,000 feet
SA-6	This missile is considerably larger than the American-built Hawk but is assumed to have the same basic low-altitude defense role. Each unit consists of a tracked vehicle with three solid propellant missiles.	Unknown
SA-7	An optically guided missile with an infrared warhead and shoulder-firing capability used against low-flying aircraft.	3,000 feet
ZSU-23-4	A self-propelled tracked vehicle system with four 23mm. guns on a common mounting, and fired together. Fire control radar computes target speed and height. Carrier vehicles same as PT-76 tank.	3,900 feet
ZSU-57-2	The same as above except it uses two 57mm. guns.	—

Table A-8 (continued)

Missile	Characteristics	Effective Ceiling
French		
Crotale	This French-produced weapon is a completely automatic, short-range missile system which is meant to be used for all-weather interception of low-altitude targets. The missile is propelled by a single-stage solid propellant motor.	9,800 feet
Mascura	Missile system carried on French frigates with either beam riding/command guidance system with proximity fused warhead (Mod 2), or semi-active homing radar on missile and warhead (Mod 3). The system also contains DRBI 23 three-dimensional surveillance radar, radar data extractor, remote control directors linked with tracker radar, target illuminator radar for the Mod 3 (all Thomson-CSF), and IBM launch computer. Speed: Mach 2.5. Range: 22 nautical miles.	Unknown
Roland	Mobile, tube-launched missile system designed for use against low-flying aircraft developed by France and Germany. Roland I has an optical, infrared command guidance and tracking system, a target acquisition and identification radar subsystem. Roland II is the same, plus automatic target tracking radar. Speed: Mach 1.6.	21,000 feet
British		
Seacat	A simple, low-cost naval SAM with optical/command guidance system and a proximity-fused warhead. It also can be used as an SSM against targets within visual range. Capable of varying degrees of sophistication such as integration with ships fire control systems.	11,400 feet
Sea Dart	A third-generation British naval SAM capable of hitting targets at both very high and low altitudes. With a semi-active homing radar system and proximity-fused high-explosive warhead. Sea Dart also has a surface-to-surface capability. Propelled by a ramjet engine, it can fly at controlled speeds. It also uses a proportional navigation law which can be changed during flight. Slant range: 18.6 miles.	Unknown
Sealug Mk-1 and Mk-2	A long-range beam riding high-explosive warhead. It also has a surface-to-surface capability. Mk-1 has a slant range of 28 miles. Mk-2 has a slant range of 36 miles.	49,000 feet

Source: *Jane's Weapon Systems 1974-75.*

Table A-9
CURRENT ARMY STRENGTHS IN THE MEDITERRANEAN

Country	Number
Albania	30,000
Algeria	55,000[a]
Cyprus	—[b]
Egypt	275,000[a]
France	331,500[a]
Greece	121,000[a]
Israel	375,000[a]
Italy	306,500
Lebanon	14,000
Libya	25,000
Malta	—[c]
Morocco	55,000[a]
Spain	220,000
Syria	150,000[a]
Tunisia	20,000
Turkey	365,000[a]
Yugoslavia	190,000[a]

[a] Indicates significant airborne and/or airmobile capability.

[b] Even before the Cyprus crisis of July 1974, there was no simple figure for the total number of armed forces in Cyprus. The figures before 15 July 1974 were:

650	Regular Turkish army
950	Regular Greek army
10,000	Greek-Cypriot National Guard
10,000	TMT (Turkish-Cypriot Resistance Organization)
2,300	United Nations peace-keeping troops

After the Turkish invasion of Cyprus on 20 July 1974, the number of regular Turkish troops rose to about 37,000. As the crisis subsided, Turkey withdrew some of its forces but continues to maintain an enlarged garrison on the island. The United Nations contingent grew to 4,300 after the cease-fire.

[c] Again no simple figure is possible. The Malta Land Force (MLF) has a 688-strong contingent trained in small arms and fire-fight tactics. The MLF also operates a small flotilla of patrol boats for anti-smuggling and immigration control. There is also the 3,200-man Pioneer Corps, a paramilitary force created by Mintoff in 1972. Internal security is handled by the 1,300-man police force.

Source: *The Military Balance 1975–76.*

Table A-10

ARMORED VEHICLES IN THE MEDITERRANEAN
(tanks, armored personnel carriers and armored cars)

Country of Possession	Type		Number
Albania	Medium tanks	T-34	70
		T-54 ⎫	
		T-59 ⎬	15
		T-62	40
	APC	BA-64 ⎫	
		BTR-40 ⎬	20
		BTR-152 ⎭	
Algeria	Medium tanks	T-34	100
	Medium tanks	T-54/55	300
	Light tanks	AMX-13	50
		BTR-152	350
	APC	BTR-40	30
		BTR-50	40
		BTR-60	20
Egypt	Heavy tanks	JS-3/T-10	25
	Medium tanks	T-54/55	1,100
	Medium tanks	T-62	820
	Light amphibious tanks	PT-76	30
		BTR-40 ⎫	
		BTR-50P	
	APC	BTR-60P ⎬	2,500
		OT-64	
		BTR-152 ⎭	
	AFV	BMP-76PB	100
France	Medium tanks	AMX-30	950
	Light tanks	AMX-13	1,120
	Heavy armored cars	Panhard EBR ⎫	
	Light armored cars	Panhard AML ⎭	620
	APC	VP-90	X
	APC	AMX-10	150
		Other AFV	330
Greece	Medium tanks	M-47	300
		M-48	500
		AMX-30 (130 on order)	60
	Light tanks	M-24 ⎫	200
	Light tanks	M-41 ⎭	
	Armored cars	M-8	X
	Armored cars	M-20	X
	APC	M-59	X
	APC	M-113	X

Table A-10 (continued)

Country of Possession	Type		Number
Israel	Medium tanks, converting to self-propelled artillery	Sherman	200
	Medium tanks	Centurion	900
		M-48 (more on order)	400
		M-60 (more on order)	450
		T-54/55	400
		Other medium tanks	550
		T-62	150[a]
	Light tanks	PT-76	65
	AFV	AML-90	15
		Other AFV (including AML-60)	3,600[a]
	Armored cars	Staghound	X
	APC	M-2 M-3 M-113 (more on order) BDRM BTR-40 BTR-50P OT-62 BTR-60P BTR-152	3,300[a]
Italy	Medium tanks	M-47	700
		M-60	300
		Leopard (500 more on order)	300
	APC	M-113 AMX	3,300
	APC	LTV-7 (on order)	—
Lebanon	Medium tanks	Charioteer	60
	Light tanks	AMX-13	25
	Light tanks	M-41	18
	Armored cars	M-706 M-6 Panhard M-3 AEC	100
	APC	M-113	80
	APC	M-59	16
Libya	Medium tanks	T-62	50
		T-54/55	280
		T-34	15
	Armored cars	Saladin	100
	Scout cars	Ferret	25
	APC	BTR-40 BTR-50 BTR-60	220
		Saracen	30
		OT-64	110
		M-113AL	170

149

Country of Possession	Type		Num
Morocco	Medium tanks	M-48	
	Medium tanks	T-54	1
	Light tanks	AMX-13	1,
	Armored cars	EBR-75	
		AML-245	
		M-8	
	APC	M-3 Halftrack	
	APC	OT-64	
Spain		AMX-30 (180 on order)	
	Medium tanks	M-47	
		M-48	3
	Light tanks	M-41	1
	Scout cars	AML-60/90	
	Scout cars	M-3	
	APC	M-113	4
Syria		T-34	1
	Medium tanks	T-54/55	1,3
		T-62	7
	Light tanks	PT-76	
	APC	BTR-50	
		BTR-60	1,1
		BTR-152	
Tunisia	Light tanks	AMX-13	
	Light tanks	M-41	
	Armored cars	Saladin	
		EBR-75	
		AML-60	
		M-8	
Turkey	Medium tanks	M-47	
	Medium tanks	M-48	1,5
	Armored cars	M-8	
	APC	M-59	
	APC	M-113	1,0
Yugoslavia		T-54/55	
	Medium tanks	T-34	1,5
		M-47	
		M-4	6
	Light tanks	PT-76	
	APC	M-3	
		M-8	
		BTR-50P	
		BTR-60P	
		BTR-152	
		M-60	

a Approximate.

X = exact number unknown.

Source: *The Military Balance 1975–76.*

Table A-11
MAJOR ANTI-TANK WEAPONS IN THE MEDITERRANEAN

Country	Type
Albania	76mm. AT 85mm. AT
Algeria	Sagger ATGW
Egypt	82mm. recoilless rifles 107mm. recoilless rifles 57mm. AT 85mm. AT 100mm. AT Sagger ATGW Swatter ATGW Snapper ATGW
France	STRIM ATGW Milan ATGW SS-11/12 ATGW Harpon ATGW HOT ATGW 57mm. recoilless rifles 75mm. recoilless rifles 105/6mm. recoilless rifles
Israel	106mm. recoilless rifles LAW ATGW TOW ATGW Cobra ATGW SS-10/11 ATGW Sagger ATGW
Italy	Mosquito ATGW Cobra ATGW SS-11 ATGW TOW ATGW 57mm. recoilless rifles 75mm. recoilless rifles 106mm. recoilless rifles
Lebanon	TOW (on order)
Libya	Vigilant ATGW
Spain	89mm. recoilless rifles 106mm. recoilless rifles 90mm. SP/AT guns 75mm. AT guns
Syria	Snapper ATGW Sagger ATGW Swatter ATGW

Table A-11 (continued)

Country	Type
Turkey	SS-11 ATGW
	Cobra ATGW
	57mm. recoilless rifles
	75mm. recoilless rifles
	106mm. recoilless rifles
Yugoslavia	M-36 tank destroyers
	82mm. recoilless rifles
	75mm. recoilless rifles
	57mm. AT
	75mm. AT
	100mm. AT
	Snapper ATGW
	Sagger ATGW

Note: AT = anti-tank.
ATGW = anti-tank guided weapon.
Source: *The Military Balance 1975–76.*

Table A-12

GROUND SURFACE-TO-SURFACE MISSILES
IN THE MEDITERRANEAN

Country of Possession	Type	Number	Comments
Algeria	Frog-4	15	
Egypt	Frog-7	18	
	SCUD	24	Up to 24
	Samlet	X	
France	Pluton	12	
	Honest John	8	Converting to Pluton by end of 1975
Greece	Honest John	8	
Israel	MD-660 Jericho	X	Believed operational 280-mile range
	Ze'ev (Wolf)	X	
Italy	Honest John	X	
	Lance	X	
Syria	Frog-7	X	
	SCUD	X	
Turkey	Honest John	12	
United States	Sergeant	X	One battalion based in Italy

X = exact number unknown.
Source: *The Military Balance 1975–76.*

Table A-13

CHARACTERISTICS OF GROUND SURFACE-TO-SURFACE MISSILES

Missile	Range (miles)	Characteristics
Sergeant	28/87	A second generation battlefield support missile with nuclear or high-explosive warhead option and inertial guidance system. System improved with use of digital computer.
Honest John	4.6/23	A mobile, unguided battlefield support missile capable of carrying either nuclear or high-explosive warhead. With no electrical controls, it is fired like conventional artillery, however from a highly mobile launcher.
Lance	68	A guided missile designed to replace Honest John and Sargeant. With a simplified inertial guidance system, it can be fired from highly mobile land launchers or from planes. Speed: supersonic.
SCUD A	50/93	Heavy artillery rocket with movable tail fins fired from a tracked vehicle. It has a radio command guidance system, although later models are believed to have a simplified inertial system. It has the option of carrying either a nuclear warhead or high explosive.
SCUD B	175	A larger version of SCUD with a simplified inertial guidance system.
Frog 7	37	A spin-stabilized unguided tactical missile with a nuclear or high-explosive option. Launched from a wheeled, erector launcher—ZIL-135. This missile is similar to the American Honest John.
Samlet	124	The Samlet is a surface-to-surface version of the air-to-surface jet-powered Kennel cruise missile. It is launched from a trailer-like platform and is believed to have a semi-active homing device. Subsonic speed.
Pluton	6/74	A surface-to-surface tactical nuclear missile fired from an AMX-30 tank chassis. It has a simplified inertial guidance system and additional command vehicles with a third-generation data processing system capable of operation under severe environmental conditions.
MD-660 Jericho	280	A battlefield support missile, believed to be in production in Israel. Its warhead could be either nuclear or high-explosive.
Ze'ev (Wolf)		A short-range SSM that comes in two versions: one with a 170kg. warhead with a range of only about 1,000 meters; the second has a 70kg. warhead and travels 4,500 meters. Used extensively during the 1973 Arab-Israeli war, the Wolf is believed not to be highly accurate. Little else is known about this missile.

Source: *Jane's Weapon Systems 1974–75.*

APPENDIX B

THE MONTREUX CONVENTION

Convention Regarding the Regime of the Straits between Bulgaria, France, Great Britain (also on Behalf of Parts of the British Empire Not Represented at the Montreux Conference), Australia, Greece, Japan, Rumania, Turkey, Union of Soviet Socialist Republics, and Yugoslavia,[1] 20 July 1936

Article 1

The High Contracting Parties recognise and affirm the principle of freedom of transit and navigation by sea in the Straits.

The exercise of this freedom shall henceforth be regulated by the provisions of the present Convention.

SECTION I

MERCHANT VESSELS

Article 2

In time of peace, merchant vessels shall enjoy complete freedom of transit and navigation in the Straits, by day and by night, under any flag and with any kind of cargo, without any formalities, except as provided in Article 3 below. No taxes or charges other than those authorised by Annex I to the present Convention shall be levied by the Turkish authorities on these vessels when passing in transit without calling at a port in the Straits.

In order to facilitate the collection of these taxes or charges merchant vessels passing through the Straits shall communicate to the officials at the stations referred to in Article 3 their name, nationality, tonnage, destination and last port of call (provenance).

Pilotage and towage remain optional.

League of Nations, "Convention Regarding the Regime of the Straits, Signed at Montreux, July 20th, 1936," *League of Nations Treaty Series*, vol. 168, Geneva, 1936, pp. 214-41.

Article 3

All ships entering the Straits by the Aegean Sea or by the Black Sea shall stop at a sanitary station near the entrance to the Straits for the purposes of the sanitary control prescribed by Turkish law within the framework of international sanitary regulations. This control, in the case of ships possessing a clean bill of health or presenting a declaration of health testifying that they do not fall within the scope of the provisions of the second paragraph of the present Article, shall be carried out by day and by night with all possible speed, and the vessels in question shall not be required to make any other stop during their passage through the Straits.

Vessels which have on board cases of plague, cholera, yellow fever, exanthematic typhus or smallpox, or which have had such cases on board during the previous seven days, and vessels which have left an infected port within less than five times twenty-four hours shall stop at the sanitary stations indicated in the preceding paragraph in order to embark such sanitary guards as the Turkish authorities may direct. No tax or charge shall be levied in respect of these sanitary guards and they shall be disembarked at a sanitary station on departure from the Straits.

Article 4

In time of war, Turkey not being belligerent, merchant vessels, under any flag or with any kind of cargo, shall enjoy freedom of transit and navigation in the Straits subject to the provisions of Articles 2 and 3.

Pilotage and towage remain optional.

Article 5

In time of war, Turkey being belligerent, merchant vessels not belonging to a country at war with Turkey shall enjoy freedom of transit and navigation in the Straits on condition that they do not in any way assist the enemy.

Such vessels shall enter the Straits by day and their transit shall be effected by the route which shall in each case be indicated by the Turkish authorities.

Article 6

Should Turkey consider herself to be threatened with imminent danger of war, the provisions of Article 2 shall nevertheless continue to be applied except that vessels must enter the Straits by day and that their transit must be effected by the route which shall, in each case, be indicated by the Turkish authorities.

Pilotage may, in this case, be made obligatory, but no charge shall be levied.

Article 7

The term "merchant vessels" applies to all vessels which are not covered by Section II of the present Convention.

SECTION II

VESSELS OF WAR

Article 8

For the purposes of the present Convention, the definitions of vessels of war and of their specification together with those relating to the calculation of tonnage shall be as set forth in Annex II to the present Convention.

Article 9

Naval auxiliary vessels specifically designed for the carriage of fuel, liquid or non-liquid, shall not be subject to the provisions of Article 13 regarding notification, nor shall they be counted for the purpose of calculating the tonnage which is subject to limitation under Articles 14 and 18, on condition that they shall pass through the Straits singly. They shall, however, continue to be on the same footing as vessels of war for the purpose of the remaining provisions governing transit.

The auxiliary vessels specified in the preceding paragraph shall only be entitled to benefit by the exceptional status therein contemplated if their armament does not include: for use against floating targets, more than two guns of a maximum calibre of 105 millimetres; for use against aerial targets, more than two guns of a maximum calibre of 75 millimetres.

Article 10

In time of peace, light surface vessels, minor war vessels and auxiliary vessels, whether belonging to Black Sea or non-Black Sea

Powers, and whatever their flag, shall enjoy freedom of transit through the Straits without any taxes or charges whatever, provided that such transit is begun during daylight and subject to the conditions laid down in Article 13 and the Articles following thereafter.

Vessels of war other than those which fall within the categories specified in the preceding paragraph shall only enjoy a right of transit under the special conditions provided by Articles 11 and 12.

Article 11

Black Sea Powers may send through the Straits capital ships of a tonnage greater than that laid down in the first paragraph of Article 14, on condition that these vessels pass through the Straits singly, escorted by not more than two destroyers.

Article 12

Black Sea Powers shall have the right to send through the Straits, for the purpose of rejoining their base, submarines constructed or purchased outside the Black Sea, provided that adequate notice of the laying down or purchase of such submarines shall have been given to Turkey.

Submarines belonging to the said Powers shall also be entitled to pass through the Straits to be repaired in dockyards outside the Black Sea on condition that detailed information on the matter is given to Turkey.

In either case, the said submarines must travel by day and on the surface, and must pass through the Straits singly.

Article 13

The transit of vessels of war through the Straits shall be preceded by a notification given to the Turkish Government through the diplomatic channel. The normal period of notice shall be eight days; but it is desirable that in the case of non-Black Sea Powers this period should be increased to fifteen days. The notification shall specify the destination, name, type and number of the vessels, as also the date of entry for the outward passage and, if necessary, for the return journey. Any change of date shall be subject to three days' notice.

Entry into the Straits for the outward passage shall take place within a period of five days from the date given in the original notification. After the expiry of this period, a new notification shall be given under the same conditions as for the original notification.

When effecting transit, the commander of the naval force shall, without being under any obligation to stop, communicate to a signal station at the entrance to the Dardanelles or the Bosphorus the exact composition of the force under his orders.

Article 14

The maximum aggregate tonnage of all foreign naval forces which may be in course of transit through the Straits shall not exceed 15,000 tons, except in the cases provided for in Article 11 and in Annex III to the present Convention.

The forces specified in the preceding paragraph shall not, however, comprise more than nine vessels.

Vessels, whether belonging to Black Sea or non-Black Sea Powers, paying visits to a port in the Straits, in accordance with the provisions of Article 17, shall not be included in this tonnage.

Neither shall vessels of war which have suffered damage during their passage through the Straits be included in this tonnage; such vessels, while undergoing repair, shall be subject to any special provisions relating to security laid down by Turkey.

Article 15

Vessels of war in transit through the Straits shall in no circumstances make use of any aircraft which they may be carrying.

Article 16

Vessels of war in transit through the Straits shall not, except in the event of damage or peril of the sea, remain therein longer than is necessary for them to effect the passage.

Article 17

Nothing in the provisions of the preceding Articles shall prevent a naval force of any tonnage or composition from paying a courtesy visit of limited duration to a port in the Straits, at the invitation of the Turkish Government. Any such force must leave the Straits by the same route as that by which it entered, unless it fulfills the conditions required for passage in transit through the Straits as laid down by Articles 10, 14 and 18.

Article 18

(1) The aggregate tonnage which non-Black Sea Powers may have in that sea in time of peace shall be limited as follows:

(*a*) Except as provided in paragraph (*b*) below, the aggregate tonnage of the said Powers shall not exceed 30,000 tons;

(*b*) If at any time the tonnage of the strongest fleet in the Black Sea shall exceed by at least 10,000 tons the tonnage of the strongest fleet in that sea at the date of the signature of the present Convention, the aggregate tonnage of 30,000 tons mentioned in paragraph (*a*) shall be increased by the same amount, up to a maximum of 45,000 tons. For this purpose, each Black Sea Power shall, in conformity with Annex IV to the present Convention, inform the Turkish Government, on the 1st January and the 1st July of each year, of the total tonnage of its fleet in the Black Sea; and the Turkish Government shall transmit this information to the other High Contracting Parties and to the Secretary-General of the League of Nations;

(*c*) The tonnage which any one non-Black Sea Power may have in the Black Sea shall be limited to two-thirds of the aggregate tonnage provided for in paragraphs (*a*) and (*b*) above;

(*d*) In the event, however, of one or more non-Black Sea Powers desiring to send naval forces into the Black Sea, for a humanitarian purpose, the said forces, which shall in no case exceed 8,000 tons altogether, shall be allowed to enter the Black Sea without having to give the notification provided for in Article 13 of the present Convention, provided an authorization is obtained from the Turkish Government in the following circumstances: if the figure of the aggregate tonnage specified in paragraphs (*a*) and (*b*) above has not been reached and will not be exceeded by the despatch of the forces which it is desired to send, the Turkish Government shall grant the said authorization within the shortest possible time after receiving the request which has been addressed to it; if the said figure has already been reached or if the despatch of the forces which it is desired to send will cause it to be exceeded, the Turkish Government will immediately inform the other Black Sea Powers of the request for authorization, and if the said Powers make no objection within twenty-four hours of having received this information, the Turkish Government shall, within forty-eight hours at the latest, inform the interested Powers of the reply which it has decided to make to their request.

Any further entry into the Black Sea of naval forces of non-Black Sea Powers shall only be effected within the available limits of the aggregate tonnage provided for in paragraphs (a) and (b) above.

(2) Vessels of war belonging to non-Black Sea Powers shall not remain in the Black Sea more than twenty-one days, whatever be the object of their presence there.

Article 19

In time of war, Turkey not being belligerent, warships shall enjoy complete freedom of transit and navigation through the Straits under the same conditions as those laid down in Articles 10 to 18.

Vessels of war belonging to belligerent Powers shall not, however, pass through the Straits except in cases arising out of the application of Article 25 of the present Convention, and in cases of assistance rendered to a State victim of aggression in virtue of a treaty of mutual assistance binding Turkey, concluded within the framework of the Covenant of the League of Nations, and registered and published in accordance with the provisions of Article 18 of the Covenant.

In the exceptional cases provided for in the preceding paragraph, the limitations laid down in Articles 10 to 18 of the present Convention shall not be applicable.

Notwithstanding the prohibition of passage laid down in paragraph 2 above, vessels of war belonging to belligerent Powers, whether they are Black Sea Powers or not, which have become separated from their bases, may return thereto.

Vessels of war belonging to belligerent Powers shall not make any capture, exercise the right of visit and search, or carry out any hostile act in the Straits.

Article 20

In time of war, Turkey being belligerent, the provisions of Articles 10 to 18 shall not be applicable; the passage of warships shall be left entirely to the discretion of the Turkish Government.

Article 21

Should Turkey consider herself to be threatened with imminent danger of war she shall have the right to apply the provisions of Article 20 of the present Convention.

Vessels which have passed through the Straits before Turkey has made use of the powers conferred upon her by the preceding paragraph, and which thus find themselves separated from their bases, may return thereto. It is, however, understood that Turkey may deny this right to vessels of war belonging to the State whose attitude has given rise to the application of the present Article.

Should the Turkish Government make use of the powers conferred by the first paragraph of the present Article, a notification to that effect shall be addressed to the High Contracting Parties and to the Secretary-General of the League of Nations.

If the Council of the League of Nations decide by a majority of two-thirds that the measures thus taken by Turkey are not justified, and if such should also be the opinion of the majority of the High Contracting Parties signatories to the present Convention, the Turkish Government undertakes to discontinue the measures in question as also any measures which may have been taken under Article 6 of the present Convention.

Article 22

Vessels of war which have on board cases of plague, cholera, yellow fever, exanthematic typhus or smallpox or which have had such cases on board within the last seven days and vessels of war which have left an infected port within less than five times twenty four hours must pass through the Straits in quarantine and apply by the means on board such prophylactic measures as are necessary in order to prevent any possibility of the Straits being infected.

SECTION III

AIRCRAFT

Article 23

In order to assure the passage of civil aircraft between the Mediterranean and the Black Sea, the Turkish Government will indicate the air routes available for this purpose, outside the forbidden zones which may be established in the Straits. Civil aircraft may use these routes provided that they give the Turkish Government, as regards occasional flights, a notification of three days, and as regards flights on regular services, a general notification of the dates of passage

The Turkish Government moreover undertake, notwithstanding any remilitarization of the Straits, to furnish the necessary facilities for the safe passage of civil aircraft authorized under the air regulations in force in Turkey to fly across Turkish territory between Europe and Asia. The route which is to be followed in the Straits zone by aircraft which have obtained an authorization shall be indicated from time to time.

SECTION IV

GENERAL PROVISIONS

Article 24

The functions of the International Commission set up under the Convention relating to the regime of the Straits of the 24th July, 1923, are hereby transferred to the Turkish Government.

The Turkish Government undertake to collect statistics and to furnish information concerning the application of Articles 11, 12, 14 and 18 of the present Convention.

They will supervise the execution of all the provisions of the present Convention relating to the passage of vessels of war through the Straits.

As soon as they have been notified of the intended passage through the Straits of a foreign naval force the Turkish Government shall inform the representatives at Angora of the High Contracting Parties of the composition of that force, its tonnage, the date fixed for its entry into the Straits, and, if necessary, the probable date of its return.

The Turkish Government shall address to the Secretary-General of the League of Nations and to the High Contracting Parties an annual report giving details regarding the movements of foreign vessels of war through the Straits and furnishing all information which may be of service to commerce and navigation, both by sea and by air, for which provision is made in the present Convention.

Article 25

Nothing in the present Convention shall prejudice the rights and obligations of Turkey, or of any of the other High Contracting Parties members of the League of Nations, arising out of the Covenant of the League of Nations.

SECTION V

Article 26

The present Convention shall be ratified as soon as possible.

The ratifications shall be deposited in the archives of the Government of the French Republic in Paris.

The Japanese Government shall be entitled to inform the Government of the French Republic through their diplomatic representative in Paris that the ratification has been given, and in that case they shall transmit the instrument of ratification as soon as possible.

A *procès-verbal* of the deposit of ratifications shall be drawn up as soon as six instruments of ratification, including that of Turkey, shall have been deposited. For this purpose the notification provided for in the preceding paragraph shall be taken as the equivalent of the deposit of an instrument of ratification.

The present Convention shall come into force on the date of the said *procès-verbal*.

The French Government will transmit to all the High Contracting Parties an authentic copy of the *procès-verbal* provided for in the preceding paragraph and of the *procès-verbaux* of the deposit of any subsequent ratifications.

Article 27

The present Convention shall, as from the date of its entry into force, be open to accession by any power signatory to the Treaty of Peace at Lausanne signed on the 24th July, 1923.

Each accession shall be notified, through the diplomatic channel, to the Government of the French Republic, and by the latter to all the High Contracting Parties.

Accessions shall come into force as from the date of notification to the French Government.

Article 28

The present Convention shall remain in force for twenty years from the date of its entry into force.

The principle of freedom of transit and navigations affirmed in Article 1 of the present Convention shall however continue without limit of time.

If, two years prior to the expiry of the said period of twenty years, no High Contracting Party shall have given notice of denuncia-

164

ר to the French Government the present Convention shall continue
force until two years after such notice shall have been given. Any
h notice shall be communicated by the French Government to the
;h Contracting Parties.

In the event of the present Convention being denounced in
ordance with the provisions of the present Article, the High Con-
:ting Parties agree to be represented at a conference for the purpose
:oncluding a new Convention.

Article 29

At the expiry of each period of five years from the date of the
ry into force of the present Convention each of the High Con-
:ting Parties shall be entitled to initiate a proposal for amending
: or more of the provisions of the present Convention.

To be valid, any request for revision formulated by one of the
;h Contracting Parties must be supported, in the case of modifica-
ns to Articles 14 or 18, by one other High Contracting Party, and,
the case of modifications to any other Article, by two other High
ntracting Parties.

Any request for revision thus supported must be notified to all
High Contracting Parties three months prior to the expiry of the
rent period of five years. This notification shall contain details
the proposed amendments and the reasons which have given rise
hem.

Should it be found impossible to reach an agreement on these
posals through the diplomatic channel, the High Contracting
ties agree to be represented at a conference to be summoned for
 purpose.

Such a conference may only take decisions by a unanimous vote,
ept as regards cases of revision involving Articles 14 and 18, for
ich a majority of three-quarters of the High Contracting Parties
ll be sufficient.

The said majority shall include three-quarters of the High Con-
cting Parties which are Black Sea Powers, including Turkey.

In witness whereof, the above-mentioned Plenipotentiaries have
ned the present Convention.

Done at Montreux the 20th July, 1936, in eleven copies, of
ich the first copy, to which the seals of the Plenipotentiaries have
n affixed, will be deposited in the archives of the Government of
 French Republic and of which the remaining copies have been
nsmitted to the signatory Powers.

The undersigned, Plenipotentiaries of Japan, declare, i
name of their Government, that the provisions of the present
vention do not in any sense modify the position of Japan as a
not a member of the League of Nations, whether in relation
Covenant of the League of Nations or in regard to treaties of n
assistance concluded within the framework of the said Cov
and that in particular Japan reserves full liberty of interpretati
regards the provisions of Articles 19 and 25 so far as they cc
that Covenant and those treaties.

NOTE: Annex I regarding taxes and charges is deleted.

ANNEX II [2]

A. STANDARD DISPLACEMENT

(1) The standard displacement of a surface vessel is th
placement of the vessel, complete, fully manned, engined
equipped ready for sea, including all armament and ammur
equipment, outfit, provisions and fresh water for crew, miscella
stores and implements of every description that are intended
carried in war, but without fuel or reserve feed water on board

(2) The standard displacement of a submarine is the st
displacement of the vessel complete (exclusive of the water in
watertight structure), fully manned, engined and equipped reac
sea, including all armament and ammunition, equipment, outfit
visions for crew, miscellaneous stores and implements of
description that are intended to be carried in war, but without
lubricating oil, fresh water or ballast water of any kind on b

(3) The word "ton" except in the expression "metric
denotes the ton of 2,240 lb. (1,016 kilos).

B. CATEGORIES

(1) *Capital Ships* are surface vessels of war belonging to o
the two following sub-categories:

(a) Surface vessels of war, other than aircraft-carriers
iliary vessels, or capital ships of sub-category (b), the sta
displacement of which exceeds 10,000 tons (10,160 metric
or which carry a gun with a calibre exceeding 8 in. (203

---

[2] The wording of the present Annex is taken from the London Naval Tre
March 25th, 1936.

(*b*) Surface vessels of war, **other** than aircraft-carriers, **the** standard displacement of which does not exceed 8,000 tons (8,128 metric tons) and which carry a gun with a calibre exceeding 8 in. (203 mm.).

(2) *Aircraft-Carriers* are surface vessels of war, whatever their displacement, designed or adapted primarily for the purpose of carrying and operating aircraft at sea. The fitting of a landing-on or flying-off deck on any vessel of war, provided such vessel has not been designed or adapted primarily for the purpose of carrying and operating aircraft at sea, shall not cause any vessel so fitted to be classified in the category of aircraft-carriers.

The category of aircraft-carriers is divided into two sub-categories as follows:

(*a*) Vessels fitted with a flight deck, from which aircraft can take off, or on which aircraft can land from the air;

(*b*) Vessels not fitted with a flight deck as described in (*a*) above.

(3) *Light Surface Vessels* are surface vessels of war other than aircraft-carriers, minor war vessels or auxiliary vessels, the standard displacement of which exceeds 100 tons (102 metric tons) and does not exceed 10,000 tons (10,160 metric tons), and which do not carry a gun with a calibre exceeding 8 in. (203 mm.).

The category of light surface vessels is divided into three sub-categories as follows:

(*a*) Vessels which carry a gun with a calibre exceeding 6.1 in. (155 mm.);

(*b*) Vessels which do not carry a gun with a calibre exceeding 6.1 in. (155 mm.) and the standard displacement of which exceeds 3,000 tons (3,048 metric tons);

(*c*) Vessels which do not carry a gun with a calibre exceeding 6.1 in. (155 mm.) and the standard displacement of which does not exceed 3,000 tons (3,048 metric tons).

(4) *Submarines* are all vessels designed to operate below the surface of the sea.

(5) *Minor War Vessels* are surface vessels of war, other than auxiliary vessels, the standard displacement of which exceeds 100 tons (102 metric tons) and does not exceed 2,000 tons (2,032 metric tons), provided they have none of the following characteristics:

(*a*) Mount a gun with a calibre exceeding 6.1 in. (155 mm.);

(*b*) Are designed or fitted to launch torpedoes;

(*c*) Are designed for a speed greater than twenty knots.

167

(6) *Auxiliary Vessels* are naval surface vessels the standard displacement of which exceeds 100 tons (102 metric tons), which are normally employed on fleet duties or as troop transports, or in some other way than as fighting ships, and which are not specifically built as fighting ships, provided they have none of the following characteristics:

(a) Mount a gun with a calibre exceeding 6.1 in. (155 mm.);

(b) Mount more than eight guns with a calibre exceeding 3 in. (76 mm.);

(c) Are designed or fitted to launch torpedoes;

(d) Are designed for protection by armour plate;

(e) Are designed for a speed greater than twenty-eight knots;

(f) Are designed or adapted primarily for operating aircraft at sea;

(g) Mount more than two aircraft-launching apparatus.

C. Over-Age

Vessels of the following categories and sub-categories shall be deemed to be "over-age" when the undermentioned number of years have elapsed since completion:

(a) Capital ships .... 26 years
(b) Aircraft-carriers .... 20 years
(c) Light surface vessels, sub-categories (a) and (b):
    (i) If laid down before 1st January, 1920 .. 16 years
    (ii) If laid down after 31st December, 1919 . 20 years
(d) Light surface vessels, sub-category (c) ..... 16 years
(e) Submarines .... 13 years

NOTE: Annex III regarding certain Japanese training ships is deleted.

ANNEX IV

1. The categories and sub-categories of vessels to be included in the calculation of the total tonnage of the Black Sea Powers provided for in Article 18 of the present Convention are the following:

Capital Ships:
    Sub-category (a);
    Sub-category (b).

Aircraft-Carriers:
    Sub-category *(a)*;
    Sub-category *(b)*.

Light Surface Vessels:
    Sub-category *(a)*;
    Sub-category *(b)*;
    Sub-category *(c)*.

Submarines:
    As defined in Annex II to the present Convention.

The displacement which is to be taken into consideration in the calculation of the total tonnage is the standard displacement as defined in Annex II. Only those vessels shall be taken into consideration which are not over-age according to the definition contained in the said Annex.

2. The notification provided for in Article 18, paragraph *(b)*, shall also include the total tonnage of vessels belonging to the categories and sub-categories mentioned in paragraph 1 of the present Annex.

PROTOCOL

At the moment of signing the Convention bearing this day's date, the undersigned Plenipotentiaries declare for their respective Governments that they accept the following provisions:

(1) Turkey may immediately remilitarize the zone of the Straits as defined in the Preamble to the said Convention.

(2) As from the 15th August, 1936, the Turkish Government shall provisionally apply the régime specified in the said Convention.

(3) The present Protocol shall enter into force as from this day's date.

Done at Montreux, the 20th July, 1936.

Cover and book design: Pat Taylor